Marco Girhard

Bacterial Cytochrome P450 Monooxygenases

Marco Girhard

Bacterial Cytochrome P450 Monooxygenases

Investigations on Redox Partners and Optimization for Whole Cell Biocatalysis

Südwestdeutscher Verlag für Hochschulschriften

Impressum/Imprint (nur für Deutschland/ only for Germany)
Bibliografische Information der Deutschen Nationalbibliothek: Die Deutsche Nationalbibliothek
verzeichnet diese Publikation in der Deutschen Nationalbibliografie; detaillierte bibliografische
Daten sind im Internet über http://dnb.d-nb.de abrufbar.

Alle in diesem Buch genannten Marken und Produktnamen unterliegen warenzeichen-, marken-
oder patentrechtlichem Schutz bzw. sind Warenzeichen oder eingetragene Warenzeichen der
jeweiligen Inhaber. Die Wiedergabe von Marken, Produktnamen, Gebrauchsnamen,
Handelsnamen, Warenbezeichnungen u.s.w. in diesem Werk berechtigt auch ohne besondere
Kennzeichnung nicht zu der Annahme, dass solche Namen im Sinne der Warenzeichen- und
Markenschutzgesetzgebung als frei zu betrachten wären und daher von jedermann benutzt
werden dürften.

Verlag: Südwestdeutscher Verlag für Hochschulschriften GmbH & Co. KG
Dudweiler Landstr. 99, 66123 Saarbrücken, Deutschland
Telefon +49 681 37 20 271-1, Telefax +49 681 37 20 271-0
Email: info@svh-verlag.de
Zugl.: Düsseldorf, HHU, Diss., 2010

Herstellung in Deutschland:
Schaltungsdienst Lange o.H.G., Berlin
Books on Demand GmbH, Norderstedt
Reha GmbH, Saarbrücken
Amazon Distribution GmbH, Leipzig
ISBN: 978-3-8381-2556-5

Imprint (only for USA, GB)
Bibliographic information published by the Deutsche Nationalbibliothek: The Deutsche
Nationalbibliothek lists this publication in the Deutsche Nationalbibliografie; detailed
bibliographic data are available in the Internet at http://dnb.d-nb.de.

Any brand names and product names mentioned in this book are subject to trademark, brand
or patent protection and are trademarks or registered trademarks of their respective holders.
The use of brand names, product names, common names, trade names, product descriptions
etc. even without a particular marking in this works is in no way to be construed to mean that
such names may be regarded as unrestricted in respect of trademark and brand protection
legislation and could thus be used by anyone.

Publisher: Südwestdeutscher Verlag für Hochschulschriften GmbH & Co. KG
Dudweiler Landstr. 99, 66123 Saarbrücken, Germany
Phone +49 681 37 20 271-1, Fax +49 681 37 20 271-0
Email: info@svh-verlag.de

Printed in the U.S.A.
Printed in the U.K. by (see last page)
ISBN: 978-3-8381-2556-5

Copyright © 2011 by the author and Südwestdeutscher Verlag für Hochschulschriften GmbH &
Co. KG and licensors
All rights reserved. Saarbrücken 2011

Vorwort

Ich bin vielen Personen zu Dank verpflichtet, ohne die die Erstellung dieser Arbeit nicht möglich gewesen wäre!

Mein ganz besonderer Dank gilt Frau Prof. Dr. Vlada Urlacher für die Überlassung des interessanten Themas, die hervorragende Betreuung, gute Arbeitsbedingungen, unermüdliches Interesse und Unterstützung, kritische Diskussionen und viele hilfreiche Tipps für meine Arbeit. Außerdem möchte ich mich für die Erstbegutachtung meiner Arbeit bedanken.

Frau Prof. Dr. Martina Pohl danke ich für die freundliche Übernahme des Koreferats.

Mein Dank gilt Herrn Prof. Dr. Rolf D. Schmid und Herrn Prof. Dr. Bernhard Hauer für Ihre Unterstützung und die hervorragenden Arbeitsbedingungen während meiner Zeit am Institut für Technische Biochemie an der Universität Stuttgart.

Herrn Dr. Akira Arisawa möchte ich stellvertretend für alle Kollegen der Firma Mercian Corporation danken, die meinen dreimonatigen Forschungsaufenthalt in Japan zu einem unvergesslichen Erlebnis gemacht haben.

Frau Prof. Dr. Rita Bernhard und Herrn Dr. Yogan Khatri danke ich für die fruchtbare Kooperation und hervorragende Zusammenarbeit mit dem Institut für Biochemie der Universität des Saarlandes.

Für die Zusammenarbeit im Sesquiterpenprojekt danke ich Herrn Dr. Jens Schrader und Herrn Dr. Dirk Holtmann (Dechema e.V.).

Danke auch an Herrn Prof. Dr. Peter Dürre und Frau Stefanie Schuster vom Institut für Mikrobiologie und Biotechnologie an der Universität Ulm für die Klonierung und Überlassung von Plasmiden mit Genen aus *C. acetobutylicum*.

Mein Dank gilt ferner allen Diplomstudenten und Forschungspraktikanten, die mich während meiner Arbeit unterstützt haben. Im einzeln waren dies: Tobias Klaus, Sumire

Honda Malca, Björn Mückschel, Svetlana Tihovsky, Marion Iwanizwa und Sebastian Kriening.

Das gute Arbeitsklima und die freundschaftliche Zusammenarbeit mit meinen Arbeitskollegen des Instituts für Biochemie, Lehrstuhl II und den ehemaligen Kollegen des Instituts für Technische Biochemie haben ebenfalls zum Gelingen meiner Arbeit beigetragen.

Für die finanzielle Unterstützung danke ich der Arbeitsgemeinschaft internationaler Forschungsgemeinschaften (AiF) im Rahmen des Forschungsvorhabens 14781 N/2 (Entwicklung eines bioelektrochemischen Verfahrens zur selektiven *in vitro* Hydroxylierung von Sesquiterpenen mit molekularbiologisch optimierten P450-Monooxygenasen), sowie der Deutschen Forschungsgemeinschaft und dem Ministerium für Wissenschaft, Forschung und Kunst des Landes Baden-Württemberg im Rahmen des Sonderforschungsbereiches 706 (Katalytische Selektivoxidationen von C–H-Bindungen mit molekularem Sauerstoff).

Zum Schluss möchte ich mich bei meinen Eltern bedanken, die mich während meines Studiums immer in jeder Hinsicht unterstützt haben, und bei meinen Freunden, die mir Rückhalt und Aufmunterung gaben, wenn es mit der Arbeit mal schlechter lief.

Table of contents

Zusammenfassung V

Abstract VII

1. **Introduction** 1
   1.1. **Industrial biocatalysis (White biotechnology)** 1
      1.1.1. Bulk and fine chemicals 1
      1.1.2. Flavors and fragrances 2
   1.2. **Cytochrome P450 monooxygenases** 4
      1.2.1. Sources, functions and nomenclature 4
      1.2.2. Structure 5
      1.2.3. Substrate binding 7
      1.2.4. Catalytic mechanism 8
      1.2.5. Catalyzed reactions 10
      1.2.6. Industrial application of P450s: Examples and limitations 12
   1.3. **Electron transfer proteins** 13
      1.3.1. Topology 13
      1.3.2. Reductases, ferredoxins and flavodoxins 16
         1.3.2.1. Reductases 16
         1.3.2.2. Ferredoxins 18
         1.3.2.3. Flavodoxins 19
   1.4. **Aim of the work** 21

2. **Results** 23
   2.1. **CYP109B1 from *Bacillus subtilis* and CYP109D1 from *Sorangium cellulosum*** 23
      2.1.1. Manuscript: Characterization of the versatile monooxygenase CYP109B1 from *Bacillus subtilis* 23
         2.1.1.1. Abstract 24
         2.1.1.2. Introduction 24
         2.1.1.3. Materials and methods 27
         2.1.1.4. Results and discussion 33
         2.1.1.5. Discussion 44
         2.1.1.6. Acknowledgements 46
         2.1.1.7. Conflict of interest 46
         2.1.1.8. References (to chapter 2.2.1) 46

2.1.1.9. Supplementary Material 51

2.1.2. Manuscript: Regioselective biooxidation of (+)-valencene by recombinant
E. coli expressing CYP109B1 from *Bacillus subtilis* in a two liquid phase system .. 54

2.1.2.1. Supplementary Material 67

2.1.3. Manuscript: Regioselective hydroxylation of norisoprenoids by
CYP109D1 from *Sorangium cellulosum* 69

2.1.3.1. Abstract 70

2.1.3.2. Introduction 70

2.1.3.3. Material and methods 72

2.1.3.4. Results 78

2.1.3.5. Discussion 86

2.1.3.6. Acknowledgments 88

2.1.3.7. References (to chapter 2.1.3) 89

2.1.3.8. Supplementary Material 94

2.2. CYP152A2 from *Clostridium acetobutylicum* 96

2.2.1. Manuscript: Cytochrome P450 monooxygenase from *Clostridium
acetobutylicum*: A new α-fatty acid hydroxylase 96

2.2.2. Manuscript: Expression, purification and characterization of two *Clostridium
acetobutylicum* flavodoxins: Potential electron transfer partners for CYP152A2 .. 103

2.2.2.1. Supplementary Material 112

3. *Discussion and Outlook* 115

3.1. Cytochrome P450 monooxygenases for biocatalysis 115

3.1.1. Screening, protein expression and purification 115

3.1.2. Substrate and product spectra 116

3.1.2.1. Substrate and product spectrum of CYP109B1 116

3.1.2.2. Substrate and product spectrum of CYP152A2 118

3.1.3. Activity reconstitution 118

3.2. Whole-cell biocatalysis 120

4. *References* 123

5. *Appendix* 137

Zusammenfassung

Die Selektivoxidation von C-H Bindungen stellt in der synthetischen organischen Chemie ein weitgehend ungelöstes Problem dar. Diese Reaktion kann hingegen von Cytochrom P450 Monooxygenasen (P450 oder CYP) bei Raumtemperatur in einem einzigen Reaktionsschritt bewerkstelligt werden, was diese Enzymklasse für industrielle Biotransformationen hochinteressant macht. Cytochrom P450 Enzyme gehören zur Familie Häm *b* enthaltender Monooxygenasen und katalysieren die Spaltung von molekularem Sauerstoff, wofür zwei Elektronen benötigt werden. Diese werden in den meisten Fällen ausgehend von den Cofaktoren NADH oder NADPH über externe Redoxproteine (Reduktasen und Flavodoxine oder Ferredoxine) auf die Hämdomäne des P450 übertragen.

Durch die Entdeckung und Charakterisierung neuer P450 mit unterschiedlichen Substratspezifitäten und Aktivitäten wird die Diversität von biotechnologisch relevanten Reaktionen bzw. Produkten erweitert. Dafür steht ein stetig wachsender Pool von derzeit mehr als 18.000 annotierten P450-Sequenzen zur Verfügung. Allerdings wurde bisher nur ein Bruchteil dieser neuen P450 funktionell exprimiert und charakterisiert. Ursächlich dafür sind unter anderem folgende Limitierungen: 1) Die Identifizierung von geeigneten Redoxpartnern, die sowohl für die Herstellung der P450-Aktivität, als auch eine effiziente Biokatalyse mit P450 essentiell notwendig sind, ist oftmals schwierig. So befinden sich zum Beispiel die Gene von physiologischen Redoxproteinen meist nicht vor- oder nachgelagert des P450-Gens im Genom. 2) Ferner ist die physiologische Funktion neuer P450 oft unbekannt, so dass keine Aussagen über deren potentielle Substrate getroffen werden können, was die Suche und Auswahl geeigneter P450-Kandidaten zeit- und arbeitsintensiv macht.

Im Rahmen dieser Arbeit wurden zwei neue bakterielle P450 Enzyme, CYP109B1 aus *Bacillus subtilis* und CYP152A2 aus *Clostridium acetobutylicum* charakterisiert, ihre oxidative Aktivität mit Hilfe von verschiedenen Redoxpartnern hergestellt, und das biotechnologische Potential beider Enzyme untersucht. CYP109B1 wurde mittels Durchmusterung einer Bibliothek von 242 rekombinant in *E. coli* exprimierten P450 aus Bakterien und Pilzen gefunden, da es als einziger Kandidat das verwendete Substrat (+)-Valencen regio- und chemoselektiv zu Nootkatol und (+)-Nootkaton oxidierte. (+)-Nootkaton ist ein Aromastoff, der natürlicherweise in lediglich geringen Mengen in Grapefruitölen vorkommt und Einsatz als Duft- und Geschmacksstoff in Parfüms und Lebensmitteln findet. Im Folgenden wurde die Produktion von (+)-Nootkaton im größeren Maßstab mit einem rekombinanten *E. coli* Stamm durchgeführt, der CYP109B1 zusammen mit (nicht-physiologischen) Redoxproteinen exprimierte. Auf Basis dieses Ganzzellbiokatalysators konnten im Zweiphasensystem mit nicht-wassermischbaren organischen Lösemitteln unter optimierten Bedingungen bis zu 15 mg l^{-1} h^{-1} Nootkatol und (+)-Nootkaton produziert werden.

Das Substrat- und Produktspektrum von CYP109B1 wurde analysiert: Zusammenfassend lässt sich sagen, dass die Oxidation „linearer" Substrate, wie zum Beispiel von Fettsäuren oder n-Alkoholen, durch CYP109B1 nicht regioselektiv verläuft. Allerdings zeigt CYP109B1 hohe Regio- und Chemoselektivität für die Oxifunktionalisierung allylischer Kohlenstoffatome von cyclischen Terpenen und Terpenoiden: So wurden $α$- und $β$-Ionon mit 100%iger Selektivität zu 3-Hydroxy-$α$-Ionon bzw. 4-Hydroxy-$β$-Ionon oxidiert. 4-Hydroxy-$β$-Ionon ist ein wichtiger Baustein für die Synthese von Carotinoiden und dem Phytohormon Abscisinsäure.

Ein weiteres P450 mit Potential für biotechnologische Anwendung das im Rahmen dieser Arbeit untersucht wurde ist CYP152A2 – eine Peroxygenase aus dem anaeroben Bakterium *C. acetobutylicum*. Die Analyse des Substratspektrums ergab, dass CYP152A2 die Oxidation von gesättigten Fettsäuren selektiv am $α$- oder $β$-Kohlenstoffatom der Alkylkette katalysiert. Derart hydroxylierte Produkte finden unter anderem Anwendung bei der Herstellung von Surfactin-Antibiotika. Ferner wurde gezeigt, dass CYP152A2 auch mit Wasserstoffperoxid katalytisch aktiv ist und somit nicht notwendigerweise Redoxproteine für die Übertragung von Elektronen benötigt. Durch den Einsatz hoher Konzentrationen von Wasserstoffperoxid konnten Reaktionsraten von bis zu 200 min^{-1} erreicht werden, jedoch wurde das Enzym unter diesen Bedingungen binnen weniger Minuten inaktiviert. Kamen Redoxproteine für die Biokatalyse mit CYP152A2 zum Einsatz, wurden niedrigere Reaktionsraten gemessen, allerdings wurde aufgrund einer höheren P450-Stabilität die Effizienz des Systems insgesamt verbessert und somit bis zu 40-fach höhere Substratumsätze erreicht.

Durch die Verwendung von nicht-physiologischen Redoxpartnern für die Biokatalyse mit CYP109B1 und CYP152A2 kam es zu Limitierungen der katalysierten Reaktionen durch die Entkopplung von NAD(P)H-Verbrauch und Substratoxidation. Daher wurden potentielle physiologische Redoxpartner für beide P450 identifiziert, die bis zu sechsfach höhere Kopplung gegenüber den zuvor verwendeten nicht-physiologischen Redoxpartnern zeigten. Hierdurch konnte die biokatalytische Aktivität von CYP152A2 und CYP109B1 deutlich gesteigert werden.

Abstract

Cytochrome P450 enzymes (P450 or CYP) are heme *b* containing monooxygenases that introduce one atom of molecular oxygen into a vast range of compounds and catalyze a broad spectrum of reactions, where chemical catalysts often fail. Oxidations by P450s require the consecutive delivery of two electrons derived from the pyridine cofactors NAD(P)H and transferred to the heme iron via external redox proteins (reductases and flavodoxins or ferredoxins). The biotechnological potential of P450s was recognized many years ago; however, the characterization and exploitation of newly-discovered P450 sequences is hampered by two major limitations: 1) The need for suitable electron transfer partners mandatory for P450 activity and for efficient biocatalysis, and 2) the unknown physiological function of most P450s meaning that there is no information available on potential substrates, which makes the selection of candidate P450s time- and labor-intensive.

Within this study the screening of a recombinant P450 library comprising 242 bacterial and fungal P450s was accomplished. The aim was to identify and characterize novel P450s whose oxidizing activities lead to high-value fine-chemicals, especially for flavor and fragrance industries. One example was (+)-nootkatone, which is a sought-after fragrance naturally found in low amounts in grapefruit. Within the screening, CYP109B1 from *Bacillus subtilis* was found to produce (+)-nootkatone by oxidation of the precursor (+)-valencene via the intermediate nootkatol. A whole cell process with recombinant *E. coli* expressing CYP109B1 in a two-liquid phase system was designed and produced 15 mg l^{-1} h^{-1} of (+)-nootkatone and nootkatol under optimized conditions. Further, the substrate and product spectrum of CYP109B1 was analyzed: The enzyme is generally unselective for fatty acid oxidation, but shows high selectivity for allylic oxidation of cyclic terpenes and terpenoids, for example 100% for α- and β-ionone.

A second P450 focused on within this study with potential for biotechnological application was CYP152A2 – a peroxygenase from the anaerobe bacterium *Clostridium acetobutylicum*. This P450 was found to catalyze fatty acid oxidation at α- and β-carbon atoms exclusively.

Non-physiological redox proteins were used for activity reconstitution of both P450s in the first experiments, but were proven inefficient due to uncoupling between NAD(P)H consumption and substrate oxidation. Hence, physiological redox partners for both P450s were identified and characterized. Their application resulted in higher coupling (to some extend) and thus led to improved biocatalytic P450 activities.

1. Introduction

1.1. Industrial biocatalysis (White biotechnology)

Biocatalysis in industrial synthetic chemistry has lately experienced significant growth. Many European chemical companies are reorganizing resulting in a general opening towards bioprocesses – so-called "White biotechnology" - or hybrid chemical/biocatalytic processes [1]. This industrial trend combined with a strong preference of consumers for natural products are driving forces for novel biotechnological solutions [2]. Enzymes are remarkable catalysts: They can accept a wide range of complex molecules as substrates, and they often catalyze reactions with high regio- and enantioselectivities. Such biocatalysts can therefore be used in both simple and complex biotransformations without the need for protecting steps that are common in organic synthesis [3].

1.1.1. Bulk and fine chemicals

Advances in enzymatic catalysis have been extended to the synthesis of fine chemicals (especially in the food and pharmaceutical industries, where high reaction selectivity on complex substrates is critical), of polymers [4], and also of some bulk chemicals, for example the hydration of acrylonitrile into acrylamide catalyzed by nitril hydratase [5]. The application of a biocatalytic process for the production of bulk chemicals, however, is not economically feasible in most cases, since it depends on several factors like the type of biocatalyst, its recovery or reuse, or the need for specific reactor and hardware configurations. Biotransformations are therefore usually used for products at a scale of up to 10,000 t/a only. Examples are represented by the production of amides, amines and alcohols (BASF), amino acids, 6-aminopenicillanic acid (DSM), or the production of heterocyclic compounds (Lonza). Commonly used enzymes in this case are easy to handle hydrolases (lipases, acylases, nitrilases or amidases) [3]. The use of other enzymes, for example oxidoreductases, is far more complicated due to cofactor dependency and generally low stability and activity (see chapter 1.2.6). Biocatalytic processes utilizing such enzymes are therefore feasible only, if a high value creation is reached, which holds true for the synthesis of some fine chemicals, for example pharmaceuticals, or flavors and fragrances for food industries.

1.1.2. Flavors and fragrances

Terpenes belong to the chemical group of isoprenoids and are the largest group of natural products encompassing over 40,000 known compounds. They have the general formula $(C_5H_8)_n$ and are synthesized from two isoprene units: isopentenyl diphosphate (IPP) and dimethylallyl diphosphate (DMAPP) (Figure 1-1). Classification is arranged according to the number of carbons; the major groups of interest here are monoterpenes (C10) and sesquiterpenes (C15) [6].

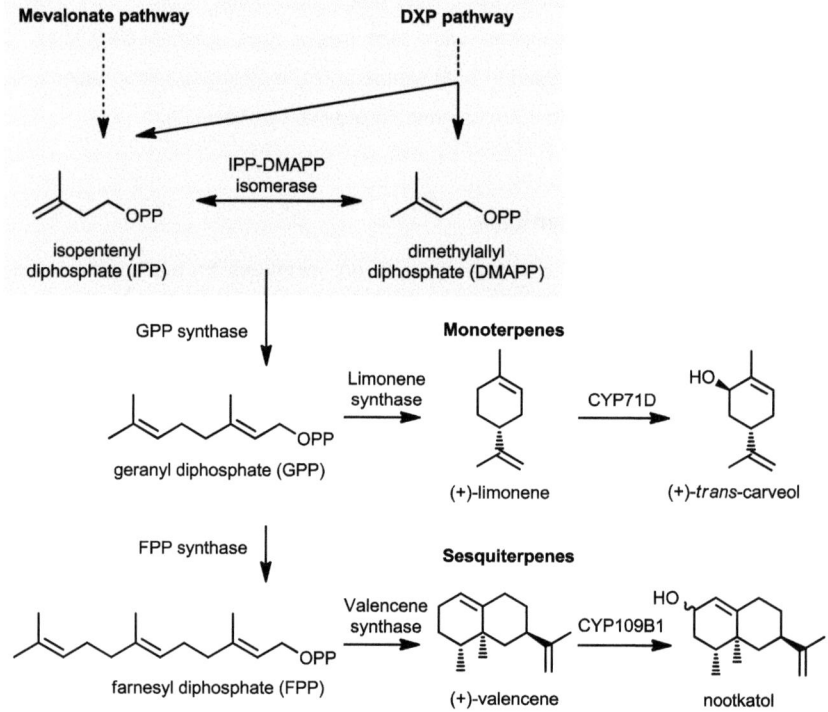

Figure 1-1: Overview of terpene pathways. The DXP and mevalonate pathways produce isopentenyl diphosphate (IPP) and dimethylallyl diphosphate (DMAPP) from central metabolites, which can be transferred into each other by IPP-DMAPP isomerase. IPP and DMAPP are converted to terpene synthase precursors. One example of a terpene synthase reaction and following reactions, e.g. oxidation by a P450, is given for each category of isoprenoids. Multiple steps are indicated by dashed lines.

Many terpene hydrocarbons are abundant in nature, for example limonene and pinene [7]. Due to their chemical instability and poor sensory impact, they are not qualified as flavorings, but represent ideal starting materials for biocatalytic oxyfunctionalization [8]. Oxygenated terpenes (terpenoids) are often sought-after pharmaceuticals or building blocks for chemical synthesis, but most important are their olfactory and gustatory characteristics. Due to this organoleptic behavior terpenoids are widely used in flavor and fragrance industries, for example for cosmetics (flowery notes) or as food additives (e.g. citric or peppermint flavors) [9].

The demand for such compounds has constantly been increasing over the past years; the estimated sales volume in 2009 was 20 billion US-$ (in comparison: 16 billion US-$ in 2005; http://www.leffingwell.com/top_10.htm; 8th September 2010). Consequently, the oxidation of low value terpenes to higher value derivatives has been recognized for a long time as an attractive opportunity for synthetic chemistry [10-11]. The demand is currently satisfied in most cases by chemical synthesis from precursors, which often requires environmentally hazardous reagents, catalysts, and solvents. Furthermore, the regiospecific introduction of carbonyl or hydroxyl groups in terpenes by chemical means has proven difficult due to similar electronic properties of the primary and secondary allylic positions, and because the allylic hydroxylation often competes with epoxidation of the corresponding C=C double bond. As a consequence, classical chemical oxidation procedures often lead to mixtures of different products. In addition, according to US and European food regulations "natural" flavor substances can only be prepared either by physical processes, for example extraction from natural sources (which usually suffers from low yields and is dependent on the availability of the raw material), or by enzymatic or microbial processes involving precursors isolated from nature. Flavors that occur in nature, but are produced by chemical synthesis can only be labeled as "nature-identical", which increasingly represents a marketing disadvantage [12]. The value of vanillin extracted from vanilla beans, for example, is calculated as being between 1200 US-$ kg^{-1} and 4000 US-$ kg^{-1}, whereas the price of synthetic vanillin prepared from guaiacol is less than 15 US-$ kg^{-1} [13]. For obvious reasons biocatalytic conversion of terpenes and terpenoids was therefore considered as early as in the 1960s [14].

A major advantage of biocatalysts is their often high chemo-, regio- and/or enantioselectivity, which is beneficial for the production of chiral products or intermediates. Chiral flavors and fragrances often occur in nature as single enantiomers, and different enantiomers or regioisomers could show different sensorial properties [15]. Most enzymes that enable regio- and enantioselective oxyfunctionalizations of terpenes belong to the group of cytochrome P450 monooxygenases.

One substrate of interest examined within this study was the sesquiterpene (+)-valencene, which can be oxidized by P450s via nootkatol to yield (+)-nootkatone, a high-priced fragrance that displays grapefruit-like taste and is widely used as flavoring in beverages and cosmetics [16] (see chapter 2.1.2 for details).

1.2. Cytochrome P450 monooxygenases

1.2.1. Sources, functions and nomenclature

Cytochrome P450 monooxygenases (P450s or CYPs) belong to the superfamily of heme b containing monooxygenases found in all domains of life [17]. They play a central role in drug metabolism and have been shown to be involved in the biosynthesis of important natural compounds like steroids, fatty acids, eicosanoids, bile acids, and the oxidation of terpenes and aromatic compounds.

The first P450s were discovered and defined as a distinct class of hemoproteins about 50 years ago [18-19]. During the next 30 years, however, the progress in identification and characterization of new P450s was slow; in 1987 the first nomenclature of P450 enzymes was proposed for 65 P450 genes known at that time. This classification was based on the amino acid identity, phylogenetic criteria and gene organization [20] and was updated in 1989 (71 P450 genes and four pseudogenes), and in 1991 (154 P450 genes and seven pseudogenes) [21-22]. Since that time, the number of known P450 sequences started to increase almost exponentially due to a vast number of genome sequencing projects. Consequently, the first P450 nomenclature was revised in 1993 [23]. According to recommendation of a nomenclature committee P450 superfamily genes were now labeled as CYP (from **Cy**tochrome **P**450). Enzymes exhibiting a sequence similarity of more than 40% belong to the same family, while P450s exhibiting more than 55% similarity form a subfamily. For example, P450$_{BM3}$ from *Bacillus megaterium* is classified as CYP102A1 meaning that it belongs to the CYP102 family, CYP102A subfamily and is identified as CYP102A1. Except of the CYP51 family, which comprises members from all five kingdoms of life, all bacterial P450s belong to families with family names > 100.

The P450 classification has constantly been updated [17, 24-25] and currently there are more than 12400 P450 sequences classified, which are available in several online databases. Examples for databases are represented by "CYPED"[1] [26-27], the "Fungal Cytochrome P450 Database"[2] listing more than 6800 fungal P450 sequences [28], or "The cytochrome P450 homepage"[3] of Dr. D. Nelson [29].

The identification speed of new P450 sequences makes it increasingly difficult to bear up with their classification and therefore up to date there are 6000 more P450 sequences known, but not yet classified [24]. Furthermore, in contrast to the increasing sequence numbers, the large oxidative potential of P450s is by now explored at a fraction only: Some very few of all annotated sequences have been cloned, and the enzymes expressed functionally and characterized so far. Two major limitations prevent the characterization of newly-discovered P450s: (i) Their physiological function is unknown in many cases meaning that there is no information available on potentially accepted substrates, and (ii) the identification of suitable electron transfer partners necessary for activity reconstitution is difficult, since for example the genes of potential physiological candidates are usually not located upstream or downstream of the P450 locus in the genome (see chapters 1.1 and 1.2.6).

1.2.2. Structure

The first structure of a P450 that was uncovered in 1985 was that of $P450_{cam}$ from *Pseudomonas putida* (CYP101) [30]. For a long time only the structures of soluble, microbial P450s were resolved, for example those of $P450_{BM3}$ [31], $P450_{terp}$ [32], $P450_{eryF}$ [33], $P450_{nor}$ [34] or $P450_{cin}$ [35]. Eukaryotic P450s are membrane-bound and therefore more difficult to crystallize. Nevertheless, in 2000 the first structure of a mammalian P450, CYP2C5 from *Oryctolagus cuniculus*, was uncovered [36], followed by the first human P450 structure of CYP2C9 in 2003 [37]. This led to great developments in the crystallization and structure determination of eukaryotic P450s. By far, thirty crystal structures of eight mammalian cytochrome P450s (CYP2C5, CYP2C8, CYP2C9, CYP3A4, CYP2D6, CYP2B4, CYP2A6 and CYP1A2) have been published [38].

Cytochrome P450 monooxygenases got their name from the unusual property to form reduced (ferrous) iron/carbon monoxide complexes in which the heme absorption Soret band shifts from 420 nm to ~ 450 nm [39]. Essential for this spectral characteristic is the axial coordination of the heme iron by a cysteine thiolate which is common to all P450s

---

[1] http://www.cyped.uni-stuttgart.de, Universität Stuttgart, 16th August 2010
[2] http://p450.riceblast.snu.ac.kr, 05th September 2010
[3] http://drnelson.uthsc.edu/CytochromeP450.html, University of Tennessee, 05th September 2010

[40-41]. The phylogenetically conserved cysteinate is the proximal ligand to the heme iron, with the distal ligand generally assumed to be a weakly bound water molecule [42-43]. Despite relatively low sequence identity across the gene superfamily, crystal structures of P450s show the same structural organization (Figure 1-2) [44].

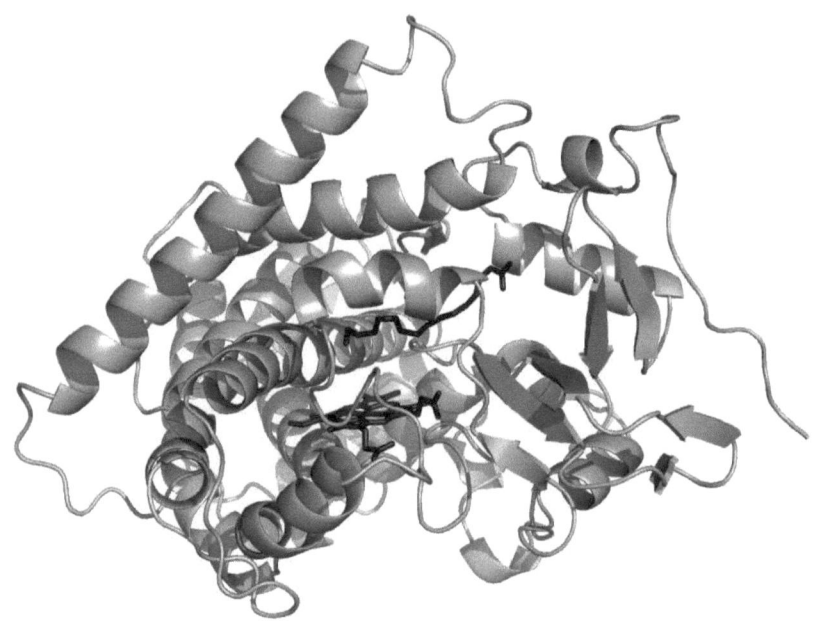

Figure 1-2: Crystal structure of the $P450_{BM3}$ monooxygenase domain from *Bacillus megaterium* with palmitoleic acid bound (adapted from pdb 1SMJ; heme and palmitoleic acid in black).

Highest structural conservation is found in the core of the protein around the heme, which reflects a common mechanism of electron- and proton-transfer, and oxygen activation. The conserved regions comprise: (i) the heme binding loop, containing the most characteristic P450 consensus sequence (Phe-Gly/Ser-X-Gly-X-His/Arg-X-Cys-XGly-X-Ile/Leu/Phe-X) with the absolutely conserved cysteine that serves as fifth ligand to the heme iron; (ii) the Glu-X-X-Arg motif, probably needed to stabilize the core structure through a salt bridge; and (iii) a consensus sequence (Ala/Gly-Gly-X-Asp/Glu-Thr), which is thought to play a role in oxygen activation through proton-transfer [45].

Besides these conserved regions, there also exist some extremely variable ones. These regions constitute the substrate-binding site that causes the acceptance of a wide range of chemically different molecules. Substrate recognition and binding is mainly arranged through six substrate recognition sites (SRS): The B' helix (SRS1), parts of helix F (SRS2), G (SRS3) and I (SRS4), as well as the β4-hairpin (SRS5) and the β2-loop (SRS6) [44]. Mutations in these regions have a high impact on substrate specificity. Crystal structures obtained from X-ray analysis of P450s with bound substrate indicate that the substrate-binding region is very flexible and often susceptible to structural reorganization upon substrate binding encouraging an induced fit model accounting for the broad substrate spectra of many P450s [46].

1.2.3. Substrate binding

Hemoproteins undergo spectrophotometrically observable transitions associated with various changes in the electronic configuration of their iron porphyrin prosthetic group. Such changes are seen on binding of substrates, acceptance or transfer of an electron, or binding and displacement of a ligand to or from the heme iron [47]. Most striking are the absorption changes which occur in the Soret region (the region of hemoprotein absorption around 400 nm) of the UV-visible spectrum linked to substrate interaction or redox changes. The absorption in this region can be attributed to (π-π*)-transitions of the porphyrin prosthetic group involving energy level changes due to electron delocalization over the conjugated double-bond system of the porphyrin ring. The position of the absorption peak in the spectrum and the magnitude of absorption are due to modifying effects of the protein, as well as to effects of any bound substrates [48].

Three different types of binding spectra have been described: (i) Type I binding is characterized by the appearance of an absorption peak at 385-390 nm and a trough at ~ 420 nm [49]; (ii) Type II is a spectral change characterized by an absorption minimum at 390-405 nm and an absorption peak at 425-435 nm; and (iii) spectral changes with an absorption maximum at 420 nm and a trough at 388-390 nm, which were originally designated as modified Type II, but were later named "reverse Type I" [50].

On the basis of the similarity between the half-maximal enzyme activity (Michaelis constant, K_M) and the half-maximal spectral change (spectral dissociation constant, K_D), it was suggested that the substrate-induced Type I spectral change represents the manifestation of an enzyme-substrate complex [49]. However, the agreement between K_M

and K_D is coincidental, since K_D is obtained from spectral changes which are determined for the ferric protein in the absence of reducing equivalents, whereas K_M is obtained from an actively metabolizing system. A comparison between K_D and K_M is therefore valid only, if substrate dissociation is far from rate limiting [48].

1.2.4. Catalytic mechanism

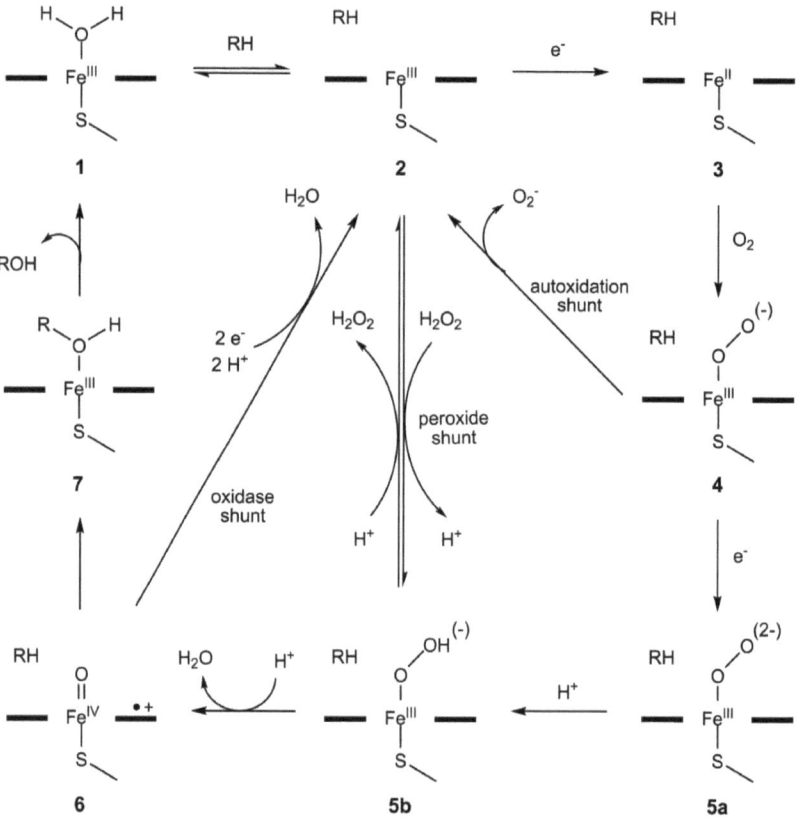

Figure 1-3: The catalytic cycle of cytochrome P450 monooxygenases.

The vast majority of cytochrome P450 monooxygenases catalyze the reductive scission of dioxygen, which requires the consecutive delivery of two electrons to the heme iron. P450s utilize reducing equivalents (electrons in form of hydrid ions) ultimately derived from the pyridine cofactors NADH or NADPH and transferred to the heme via special redox proteins. The catalytic cycle of P450s is by now well studied (Figure 1-3) [51]. Substrate binding in the active site induces the dissociation of a water molecule that is bound relatively weak as the sixth coordinating ligand to the iron (**1**), thereby inducing a shift of the heme iron spin state from low-spin to high-spin along with a positive shift in the reduction potential in the order of 130-140 mV [52]. The increased potential allows the delivery of the first electron, which reduces the heme iron from the ferric (Fe^{III}) form (**2**) to the ferrous (Fe^{II}) form (**3**). After the first electron transfer, the Fe^{II} binds dioxygen resulting in a ferrous superoxo-complex (**4**). The consecutive delivery of the second electron converts this species into a ferric peroxo anion (**5a**). This species is then protonated to a ferric hydroperoxy-complex (**5b; compound 0**). Next, protonation of the ferric hydroperoxy-complex results in a high-valent ferryl-oxo-complex (**6; compound I**) accompanied by the release of a water molecule through heterolytic scission of the dioxygen bond in the preceding intermediate (**7**).

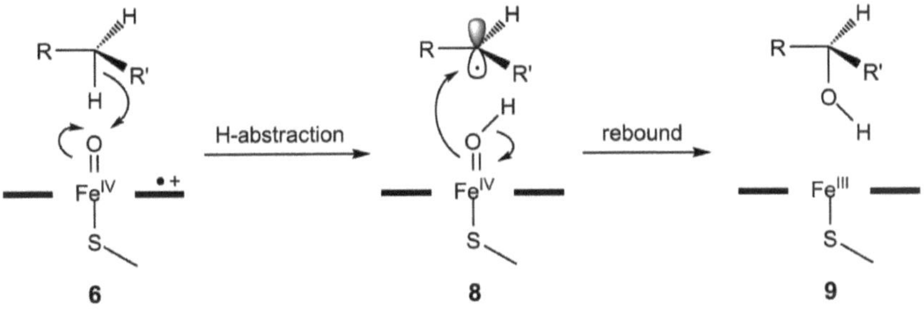

Figure 1-4: Rebound mechanism of cytochrome P450 monooxygenases.

The exact mechanistic details of oxygen insertion into the C-H bond are still subject of discussion, although it is by now widely assumed that compound I (**6**) is the oxygenating species that transfers the activated oxygen atom to the substrate. The most popular proposal is the so-called "rebound mechanism", where oxygen insertion occurs through abstraction of one hydrogen atom from the substrate to give a radical intermediate (**8**) followed by oxygen rebound to form C-OH (**9**) as shown in Figure 1-4 [53-54]. The results from numerous studies of kinetics, stereoselectivity and isotope effects for the hydroxylation reactions catalyzed by P450s conform to this proposed mechanism [55]. Alternative proposals suggest that other oxy-intermediates, such as peroxo-iron (**5a**), hydroperoxo-iron (**5b**) or H_2O_2-coordinated iron, may also be involved in substrate oxidation [56-58]. Compound 0 (**5b**), for example, was associated with the epoxidation of C=C double bonds [59-60].

The rebound mechanism is not unique to P450s, but is commonly utilized for oxygen insertion by several oxygenases, for example soluble methane monooxygenases [61-63], or peptidylglycine α-amidating monooxygenase and dopamine β-monooxygenase [64].

Under certain conditions P450s can also enter one of three so-called uncoupling pathways (see **Figure 1-3**). The "autoxidation shunt" occurs if the second electron is not delivered to reduce the ferrous superoxy-complex (**4**), which can decay forming superoxide. The inappropriate positioning of the substrate in the active site is often the molecular reason for the two other uncoupling cycles: The ferric hydroperoxy-complex (**5b**) can collapse and release hydrogen peroxide ("peroxide shunt"), while decay of compound I (**6**) is accompanied by the release of water ("oxidase shunt"). For industrial applications of P450s it is particularly important to note that the uncoupling pathways in all cases consume reducing equivalents from NAD(P)H without product formation.

It is also notable that the peroxide shunt in some cases can also be utilized by P450s to incorporate oxygen from H_2O_2 or other organic peroxides (e.g. cumene hydroperoxide or *tert*-butyl hydroperoxide) as side activity [65-68]. Furthermore, there also exists the class of natural P450 peroxygenases that can employ the peroxide shunt for catalysis and therefore do not necessarily require external redox proteins [69]. Three enzymes with a potential for biocatalytic applications are CYP152B1 (P450$_{SPα}$) from *Sphingomonas paucimobilis* [70], CYP152A1 (P450$_{Bsβ}$) from *Bacillus subtilis* [71] and CYP152A2 (P450$_{CLA}$) from *Clostridium acetobutylicum* (see chapter 2.2).

1.2.5. Catalyzed reactions

P450s are able to catalyze more than 20 different reaction types [72] including hydroxylation of non-activated sp^3 hybridized carbon-atoms, epoxidation, aromatic hydroxylation, C-C bond cleavage, heteroatom oxygenation, heteroatom release (dealkylation), oxidative ester cleavage, oxidative phenol- and ring-coupling, isomerization via (abortive) oxidation, oxidative dehalogination, as well as other complex reactions like dimer formation via Diels-Alder reactions of products or Baeyer-Villiger-type oxidations (Figure 1-5). Due to this diversity of catalyzed reactions, thousands of chemically different substrates have been described for P450s, which have been discussed and summarized in various reviews [73-77].

Hydroxylation of non-activated sp^3 hybridized carbon-atoms is probably the most important reaction catalyzed by P450s. To the first described examples of this reaction belong the hydroxylation of saturated fatty acids to hydroxy fatty acids catalyzed for example by members of the bacterial CYP102 enzymes, for example P450$_{BM3}$ [78-79], as well as the stereospecific hydroxylation of D-(+)-camphor to 5-*exo*-hydroxycamphor by P450$_{cam}$ [80].

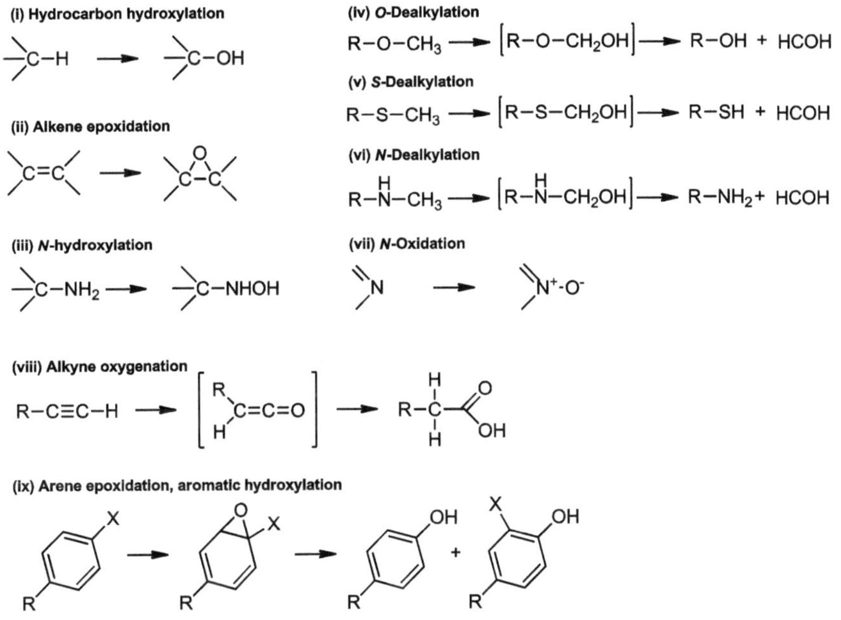

Figure 1-5: Schematic summary of the most common P450-catalyzed reactions.

Epoxidation of C=C double bonds represents another major reaction type catalyzed by P450s. A particularly attractive example in this regard is represented by P450$_{EpoK}$ from *Sorangium cellulosum*, which catalyzes the epoxidation of the macrolactone epothilone D to epothilone B – an anti-tumor agent [81].

Finally, aromatic hydroxylation also belongs to the common P450 reactions. P450$_{NikF}$ from *Streptomyces tendae* Tü901 has been described to catalyze hydroxylation of pyridylhomothreonine to form hydroxypyridylhomothreonine in the biosynthesis of nikkomycin, an inhibitor of chitin synthase [82]. Another example is the hydroxylation of naphthalene to 2-naphtol and/or 1-naphtol [74]. Many P450s have been engineered towards aromatic hydroxylation, since this ability makes them attractive candidates for the production of fine chemicals.

1.2.6. Industrial application of P450s: Examples and limitations

Cytochrome P450 monooxygenases have been studied as biocatalysts for oxidations in drug development, bioremediation and for the synthesis of fine chemicals [73, 77, 83-89]. Due to their broad distribution in nature, diverse substrate spectra, as well as established methods to change these substrate spectra, P450s are suitable candidates for the use in biotechnological applications. However, due to the cofactor dependence of P450s, their industrial applications have so far been restricted to whole-cell processes with non-recombinant microorganisms, which take advantage of the host's endogenous cofactor regeneration systems. In such instances, however, physiological effects like limited substrate uptake, toxicity of substrate or product, product degradation and elaborate downstream processing must be taken into account [90].

Microbial oxidations of steroids represent well-established large-scale commercial applications of P450s. The 11-β-hydroxylation of 11-deoxycortisol to hydrocortisone using a P450 from *Curvularia* sp. (Figure 1-6) was applied by Schering AG (in 2006 acquired by Merck, Germany) at an industrial scale of approximately 100 t/a [91]. Another example is the regioselective hydroxylation of progesterone to 11-α-hydroxyprogesterone by *Rhizopus* sp. developed in the 1950s by Pharmacia & Upjohn (later acquired by Pfizer Inc, USA) [92-93]. Both processes are one-step biotransformations, which cannot be achieved by chemical routes.

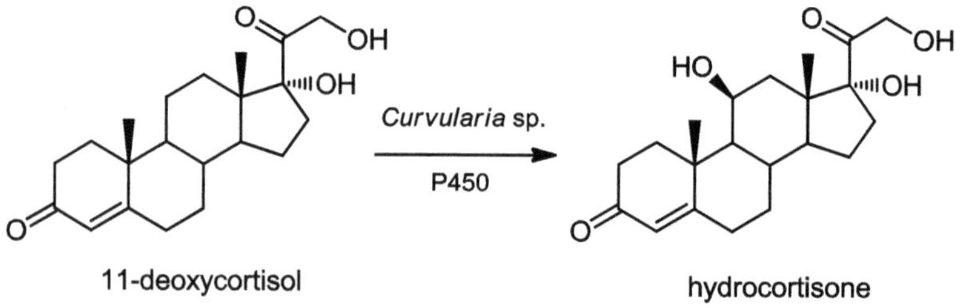

Figure 1-6: 11-β-hydroxylation of 11-deoxycortisol to hydrocortisone by *Curvularia* spec. as example for microbial oxidations of steroids at industrial scale.

Production of the cholesterol reducing pravastatin by oxidation of compactin catalyzed by a P450 from *Mucor hiemalis* (Daiichi Sankyo Inc., USA, and Bristol-Myers Squibb, USA) is another example for commercial application of microbial oxidations [94-95].

During the last decade, considerable progress has been made for recombinant expression of P450s in the well-studied hosts *Escherichia coli*, *Pseudomonas putida* and yeasts *Saccharomyces cerevisiae* and *Pichia pastoris*, which facilitates their use as industrial biocatalysts [96-101]. However, besides the challenges concerning process stability and activity of P450s, which holds true for all industrial biocatalysts, the development of technical applications for P450s faces specific problems: Probably the most important drawback that retards industrial applications is the fact that nearly all P450s require costly cofactors NADH or NADPH, which makes their application impossible if the cofactor has to be added in stoichiometric amounts. Closely linked to this cofactor dependency is the challenge to find suitable redox proteins that can sufficiently deliver electrons to the heme and to construct auxiliary redox modules. In most cases, their genes are not associated upstream or downstream of the P450 locus in the genome, which makes it difficult to reconstitute the activity of new P450s. Furthermore, interaction of a P450 with its corresponding electron donors is a necessary prerequisite of the catalytic cycle. Its specificity guarantees a sufficient reaction rate of catalysis and likewise discrimination between different potential donors and acceptors of electrons to protect the system from "shunt reactions" (see chapter 1.2.4) leading to uncoupling between NAD(P)H consumption and substrate oxidation, formation of reactive oxygen species (like hydrogen peroxide), and finally to enzyme inactivation.

1.3. Electron transfer proteins

1.3.1. Topology

As seen from the P450 catalytic cycle, the process requires the consecutive delivery of two electrons to the heme iron. P450s utilize reducing equivalents (electrons in form of hydrid ions) ultimately derived from the pyridine cofactors NADH or NADPH and transferred to the heme via special redox proteins.

Depending on the topology of the protein components involved in electron transfer, P450s can be classified. The traditional classification scheme distinguished two main classes of P450s: the class I bacterial/mitochondrial type and the class II microsomal type [73] (Figure 1-7). Class II microsomal P450s are membrane bound and accept electrons from a microsomal NADPH-cytochrome P450 reductase (CPR) containing flavin adenine dinucleotide (FAD) and flavin mononucleotide (FMN). Most of the bacterial and mitochondrial P450s belong to class I enzymes, which obtain electrons from an NAD(P)H-dependent FAD-containing reductase via an iron–sulphur protein of the [2Fe–2S] type. In bacteria, all three proteins are soluble, whereas in eukaryotes only the ferredoxin is a soluble protein of the mitochondrial matrix, while the reductase and the P450 are membrane-associated and membrane-bound to the inner mitochondrial membrane, respectively.

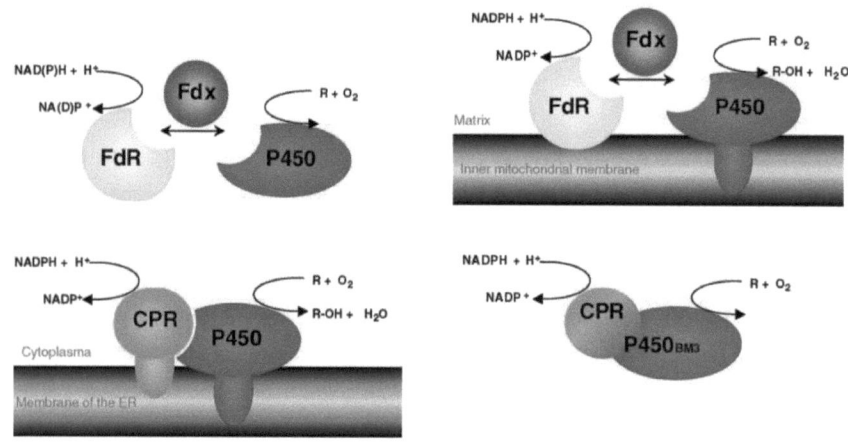

Figure 1-7: Schematic organization of different cytochrome P450 systems. Upper row, left: class I bacterial system, right: class I mitochondrial system. Lower row, left: class II microsomal system, right: self-sufficient CYP102 family (reprinted from reference [73] with permission from Elsevier).

With an increasing number of P450 sequences available, the number of described electron transfer systems has also been increasing and many of the newly-described P450 systems belonged neither to class I nor to class II. Consequently there was the need for an updated classification system taking care of those systems, which is shown in Table 1-1 [102]. This classification comprises ten different P450 classes; although some of them contain – strictly speaking – one or two members only. For example, class IX encloses up to now only the fungal nitric oxide reductase (P450$_{nor}$, CYP55) from *Fusarium oxysporum*, which is a self-sufficient P450 and therefore independent on other electron transfer proteins [103].

Table 1-1: Classes of P450 systems arranged by the topology of the protein components involved in electron transfer (reprinted from reference [102] with permission from Elsevier).

Class / Source	Electron transport chain	Localization / Example
Class I		
Bacterial	NAD(P)H > [FdR] > [Fdx]^a > [P450]	Cytosolic, soluble
Mitochondrial	NADPH > [FdR] > [Fdx] > [P450]	P450: inner mitochondrial membrane
		FdR: membrane associated
		Fdx: mitochondrial matrix, soluble
Class II		
Bacterial	NADH > [CPR] > [P450]	Cytosolic, soluble
Microsomal A	NADPH > [CPR] > [P450]	Membrane anchored, ER
Microsomal B	NADPH > [CPR] [cytb5] > [P450]	Membrane anchored, ER
Microsomal C	NADPH > [cytb5Red] [cytb5] > [P450]	Membrane anchored, ER
Class III		
Bacterial	NAD(P)H > [FdR] > [Fldx] > [P450]	Cytosolic, soluble / P450$_{cin}$
Class IV		
Bacterial	Pyruvat, CoA > [OFOR] > [Fdx] > [P450]	Cytosolic, soluble / CYP119
Class V		
Bacterial	NADH > [FdR] > [Fdx-P450]	Cytosolic, soluble
Class VI		
Bacterial	NADH > [FdR] > [Fldx-P450]	Cytosolic, soluble
Class VII		
Bacterial	NADH > [PFOR-P450]	Cytosolic, soluble
Class VIII		
Bacterial, Fungal	NADPH > [CPR-P450]	Cytosolic, soluble / P450$_{BM3}$
Class IX		
Fungal	NADH > [P450]	Cytosolic, soluble / P450$_{nor}$
Class X		
Plants, mammals	[P450]	Membrane-bound, ER

Abbreviated protein components contain the following redox centres: Fdx, ferredoxin (iron-sulfur-cluster); FdR, ferredoxin reductase (FAD); CPR, cytochrome P450 reductase (FAD, FMN); Fldx, flavodoxin (FMN); OFOR, 2-oxoacid:ferredoxin oxidoreductase (thiamine pyrophosphate, [4Fe-4S] cluster); PFOR, phthalate-family oxygenase reductase (FMN, [2Fe-2S] cluster)
^{a} Fdx containing iron-sulfur-cluster of [2Fe-2S], [3Fe-4S], [4Fe-4S], [3Fe-4S]/[4Fe-4S] type

For class IV only the soluble CYP119 from the acidothermophilic archaeon *Sulfolobus solfataricus* has been identified so far [104]. Since it was the first discovered thermophilic P450, it has been characterized intensively. Interestingly, in this case the reconstituted catalytic systems have been tested with lauric acid, because the physiological substrate for CYP119 is not yet known [105].

Class III of soluble bacterial enzymes accept electrons via a soluble FAD-containing reductase and a soluble FMN containing flavodoxin. This class comprises only P450$_{cin}$ (CYP176A) from *Citrobacter braakii* so far [35, 106]. P450$_{Biol}$ (CYP107H1) from *B. subtilis* was also reported to accept electrons utilizing either a redox system involving a flavodoxin from *C. braakii* (cindoxin) as the mediator of electron transfer [107], or the flavodoxins YkuN or YkuP from *B. subtilis* (see chapter 1.3.2.3) in combination with the (non-physiological) FAD-containing reductase from *E. coli* [108]. However, it cannot be stated that CYP107H1 belongs to class III, since there is other data available demonstrating that this enzyme also accepts electrons via a [4Fe-4S]-ferredoxin [109].

1.3.2. Reductases, ferredoxins and flavodoxins

Reductases catalyze the first step in P450 catalysis: The oxidation of the cofactor NAD(P)H to NAD(P)$^+$. In the next steps, the electrons are transferred via ferredoxins or flavodoxins, which function as (mobile) electron shuttles, to the heme-domain of the P450. This chapter provides an overview of the reductases and ferredoxins or flavodoxins that were utilized to reconstitute the activity of the P450s characterized within this study.

1.3.2.1. Reductases

Flavodoxin reductase from E. coli (FdR)

FdR is an FAD-containing oxidoreductase that transports electrons between flavodoxin or ferredoxin and NADPH. The crystal structure of FdR has been refined at 1.7 Å [115]. The enzyme is monomeric and contains one β-sandwich FAD-domain and an α/β-NADP-domain. The overall structure is similar to other reductases of the NADP-ferredoxin reductase family. FdR has been overexpressed in *E. coli* [116]. The molecular mass was determined with 27.6 kDa and the enzyme was shown to reduce cytochrome c with a k_{cat} of 141 min^{-1} and a K_M of 18 µM.

Due to its ability to transfer electrons to a large variety of flavodoxins and ferredoxins, including non-physiological redox partners, FdR is often used for activity reconstitution of P450s (see chapter 1.3.2.3).

Putidaredoxin reductase from P. putida (PdR)

PdR belongs to the superfamily of FAD-dependent pyridine nucleotide reductases. It utilizes NADH and transfers electrons to putidaredoxin (Pdx), a [2Fe-2S] ferredoxin. The crystal structure of recombinant PdR has been determined to 1.9 Å resolution [117]. The protein has a fold similar to that of disulfide reductases. Recently, the 2.6 Å X-ray structure of a catalytically active complex between PdR and Pdx, chemically cross-linked via Lys^{409PdR}-Glu^{72Pdx}, was reported [118]. Analysis of this complex allowed prediction of a 12-Å-long electron transfer route between FAD and [2Fe-2S]. Further, the surface residues Arg^{65} and Arg^{310} in PdR were proven critical for interaction with Pdx by mutation studies [119].

Bovine adrenodoxin reductase (AdR)

AdR is an FAD-containing enzyme that is a part of mitochondrial monooxygenase systems. It is membrane associated and catalyzes the electron transfer from NADPH to the electron carrier adrenodoxin (Adx), a [2Fe-2S] ferredoxin [120-121]. The crystal structure of bovine AdR has been solved by X-ray diffraction at 1.7 Å [122]. It consists of two domains with similar chain topologies, one for binding the prosthetic group FAD while the other had been suggested to bind the cofactor NADPH/NADP$^+$ [123]. AdR does not transfer two-electron packages, but belongs to the group of electron transferases that subdivide the two-electron packages from NAD(P)$^+$ into single electrons [124].

NADPH-cytochrome P450 reductase (CPR)

CPR and cytochrome P450$_{BM3}$ carry domains homologous to flavodoxin reductase and flavodoxin on a single polypeptide chain. It has therefore been hypothesized that CPR arose from a gene fusion of flavodoxin reductase and flavodoxin [110]. The reductase domain of P450$_{BM3}$ (BMR; FAD-FMN-binding domain) was cloned and the properties of the FAD- and FMN-binding subdomains were investigated [111]. Furthermore, the fatty acid monooxygenase activity of P450$_{BM3}$ was reconstituted utilizing recombinantly produced BMR [112] and its hemoprotein counterpart (BMP; heme-binding domian). In this case, the combination of BMR and BMP in a ratio of 20:1 resulted in the transfer of 80% of the reducing equivalents from NADPH for the hydroxylation of palmitic acid [113]. In later studies combinations of the separate FAD-binding domain, together with the heme-FMN-binding domain (BMP/FMN) were investigated for activity reconstitution. In this case, however, the maximal rate of oxidation of palmitic acid reached only ~ 5% of the activity of the holoenzyme [114].

1.3.2.2. Ferredoxins

General features

Ferredoxins are small, soluble electron transfer proteins that contain at least one iron-sulfur-cluster as prosthetic group and usually act as electron transfer shuttles [125]. Structurally, there exist several types of iron-sulfur proteins, which allows a classification depending on the number and type of the iron-sulfur-clusters: [2Fe-2S], [3Fe-4S], [4Fe-4S], [3Fe-4S] / [4Fe-4S], and [4Fe-4S] / [4Fe-4S] [126]. Probably the best studied enzymes of the [2Fe-2S] ferredoxins are the mammalian mitochondrial adrenodoxin and the bacterial putidaredoxin, which both are involved in class I cytochrome P450 systems (see chapter 1.3.1).

Putidaredoxin from P. putida (Pdx)

Pdx belongs to the so called "adrenodoxin-type ferredoxins" that posses a [2Fe-2S] cluster [127]. The enzyme serves as electron carrier between PdR and soluble P450s like $P450_{cam}$ (CYP101) from *P. putida*. By steady-state kinetic investigations it was shown that the k_{cat}/K_M of $P450_{cam}$ is independent on the PdR concentration and hyperbolically dependent on the Pdx concentration. Furthermore it was demonstrated that either the reduction of Pdx by PdR (at high concentrations of $P450_{cam}$), or the reduction of $P450_{cam}$ by Pdx (at high concentrations of PdR) determines the oxidation rate [128].

Several fusion proteins of PdR-Pdx-$P450_{cam}$, in which the order of the three protein domains and the linkers between them were varied, were expressed in *E. coli* and characterized [129]. The highest activity (k_{cat} = 30 min^{-1}) was obtained with a PdR-Pdx-$P450_{cam}$ construct in which short peptides of seven and four amino acids were used to link the PdR to the Pdx and the Pdx to the $P450_{cam}$ domain, respectively. Oxygen- and NADH-consumption was tightly coupled to substrate oxidation in these fusion proteins, and the rate- limiting step in the catalytic turnover was determined to be the electron transfer from Pdx to $P450_{cam}$, which was demonstrated by an increase in the activity of the $P450_{cam}$ domain upon addition of exogenous Pdx. *E. coli* cells expressing the triple fusion protein could oxidize camphor to 5-exo-hydroxycamphor and 5-oxocamphor [129], and limonene to (-)-perillyl alcohol [130].

Adrenodoxin (Adx)

In mitochondrial monooxygenase systems, the [2Fe-2S] protein Adx functions as a soluble electron carrier, which donates electrons one-at-a-time to adrenal cortex mitochondrial P450s. It first forms a complex with NADPH-reduced AdR from which it accepts one electron, dissociates, and associates with the membrane-bound P450, where it donates the electron [121]. The X-ray crystal structure of full-length oxidized Adx has been determined at 2.5 Å resolution and suggests that it exists as a dimer *in vivo* [131]. A detailed review on Adx has very recently been provided by Bernhardt and coworkers [126].

1.3.2.3. Flavodoxins

General features

Flavodoxins are small (140 to 180 residues), soluble electron transfer proteins that contain one molecule of non-covalently but tightly bound FMN as the redox active component. They were discovered in the 1960s in cyanobacteria [132-133] and clostridia [134] growing in low-iron conditions, where they replaced the iron-containing ferredoxins in reactions leading to $NADP^+$ and N_2-reduction. From sequence alignments and structural considerations, flavodoxins can be divided into two groups, short-chain and long-chain flavodoxins, that differ in the presence of twenty amino acids forming a loop with a so far unknown function [135]. Flavodoxins are involved in a variety of reactions and in some organisms they are essential, constitutive proteins. They serve for example as electron donors in the reductive activation of anaerobic ribonucleotide reductase, biotin synthase, pyruvate formate lyase, and cobalamin-dependent methionine synthase [136].

Flavodoxin from E. coli (Fdx)

Fdx is a small ferredoxin with a molecular mass of approx. 19.6 kDa. Detailed biochemical characterization of Fdx revealed that the enzyme acts as a single electron shuttle from the semiquinone form in its support of cellular functions. Furthermore, it was shown that the cytochrome c reduction rate of FdR is increased sixfold upon addition of Fdx [116].

In addition, FdR (in combination with Fdx or other flavodoxins, for example cindoxin from *C. braakii* [137]) can serve as an electron-transfer system for non-physiological microsomal and bacterial P450s, in order to substitute for unknown physiological redox partners. Some (recent) examples include CYP107H1 from *B. subtilis* [107], thermostable CYP175A1 from *Thermus thermophilus* [138], or the fatty acid hydroxylating cytochromes CYP109B1 from *B. subtilis* and CYP152A2 from *C. acetobutylicum* (see chapters 2.1 and 2.2).

Furthermore, FdR and Fdx also represent a useful model system for eukaryotic P450 reductases, because of their structural similarity to the functional domains of CPR, and have therefore been used to support activity of eukaryotic P450s, for example bovine P450c17 [139].

YkuN and YkuP from B. subtilis

YkuN and YkuP belong to the group of FMN-containing short-chain flavodoxins (158 and 151 amino acids, respectively). Both enzymes have been cloned, purified and characterized biochemically [108]. Optical and fluorimetric titrations with the oxidized flavodoxins revealed strong affinity (K_D values < 5 µM) for their potential redox partner $P450_{BioI}$. Further, stopped-flow reduction studies indicated that the maximal electron-transfer rate to fatty acid-bound $P450_{BioI}$ from YkuN and YkuP occurs considerably faster than from *E. coli* Fdx. Steady-state turnover with YkuN or YkuP, $P450_{BioI}$ and *E. coli* FdR demonstrated that both flavodoxins support hydroxylation of lauric acid with turnover rates of approx. 100 min^{-1}. Interprotein electron transfer was suggested a likely rate-limiting step in this case [108].

1.4. Aim of the work

The number of identified P450-sequences is constantly increasing as a result of rapid and automated genome sequencing. Due to their broad distribution in nature, diverse substrate spectra, as well as established methods to change these substrate spectra, cytochrome P450 monooxygenases are suitable candidates for the use in biotechnological applications. However, because of the drawbacks of missing knowledge on physiological redox partners and unclear physiological function of most P450s (described in the previous chapters), which makes it difficult to search for potential candidate P450s, examples for industrial process implementations of P450s are comparatively rare and limited to whole-cell processes with non-recombinant microorganisms so far.

The scope of this work was the screening, identification and characterization of novel regioselective P450s whose oxidizing activities lead to sought-after fine-chemicals. Two interesting candidate P450s were identified and their substrate and product spectra were analyzed: The first identified candidate was CYP109B1, a versatile monooxygenase from *Bacillus subtilis* with high regioselectivity for allylic oxidations. A whole-cell process for production of the high-priced fragrance (+)-nootkatone with recombinant *E. coli* expressing CYP109B1 in a two-liquid phase system was designed and achieved a productivity of up to 15 mg l^{-1} h^{-1}. The second candidate P450 with potential for biotechnological application was CYP152A2, a peroxygenase from *Clostridium acetobutylicum* capable of regioselective hydroxylation of fatty acids at α- and β-position.

Investigations of the electron transport systems, which are closely linked with the aim to identify suitable (physiological) redox partners for efficient biocatalysis with the newly-identified P450s were carried out. Detailed exploration of uncoupling between NAD(P)H consumption and substrate oxidation during the individual steps of the electron transport chains were undertaken aiming to increase the biocatalytic efficiency of these reconstituted systems.

2. Results

2.1. CYP109B1 from *Bacillus subtilis* and CYP109D1 from *Sorangium cellulosum*

2.1.1. Manuscript: Characterization of the versatile monooxygenase CYP109B1 from *Bacillus subtilis*

Material from this chapter appears in:

Marco Girhard[a], Tobias Klaus[a], Yogan Khatri[b], Rita Bernhardt[b] and Vlada B. Urlacher[a,*], 2010, Characterization of the versatile monooxygenase CYP109B1 from *Bacillus subtilis*, *Applied Microbiology and Biotechnology*, 87(2):595-607.

[a] Institute of Technical Biochemistry, Universität Stuttgart, 70569 Stuttgart, Germany
[b] Institute of Biochemistry, Saarland University, 66041 Saarbrücken, Germany
* Corresponding author

Material is reprinted by permission of Springer; the original manuscript is available online at: http://www.springerlink.com

2.1.1.1. Abstract

The oxidizing-activity of CYP109B1 from *Bacillus subtilis* was reconstituted *in vitro* with various artificial redox proteins including putidaredoxin reductase and putidaredoxin from *Pseudomonas putida*, truncated bovine adrenodoxin reductase and adrenodoxin, flavodoxin reductase and flavodoxin from *E. coli*, and two flavodoxins from *B. subtilis* (YkuN and YkuP). Binding and oxidation of a broad range of chemically different substrates (fatty acids, *n*-alkanes, primary *n*-alcohols, terpenoids like (+)-valencene, α- and β-ionone, and the steroid testosterone) was investigated. CYP109B1was found to oxidize saturated fatty acids (conversion up to 99%) and their methyl- and ethylesters (conversion up to 80%) at subterminal positions with a preference for the carbon atoms C11 and C12 counted from the carboxyl group. For the hydroxylation of primary *n*-alcohols the ω_{-2} position was preferred. *n*-Alkanes were not accepted as substrates by CYP109B1. Regioselective hydroxylation of terpenoides α-ionone (~ 70% conversion) and β-ionone (~ 91% conversion) yielded the allylic alcohols 3-hydroxy-α-ionone and 4-hydroxy-β-ionone respectively. Furthermore, indole was demonstrated to inhibit fatty acid oxidation.

2.1.1.2. Introduction

Cytochromes P450 (P450s or CYPs) belong to an ever-growing superfamily of heme *b* containing monooxygenases found in all domains of life (Nelson 2006). They play a central role in drug metabolism and are involved in the biosynthesis of important natural compounds. Currently there are more than 11800 P450 sequences available in several online databases, for example in CYPED (http://www.cyped.uni-stuttgart.de, Institute of Technical Biochemistry, Universitaet Stuttgart) (Fischer et al. 2007; Sirim et al. 2009) or the P450 homepage maintained by Dr. Nelson (http://drnelson.uthsc.edu/CytochromeP450.html, Molecular Sciences, University of Tennessee) (Nelson 2006). P450s catalyze the introduction of one atom of molecular oxygen in organic molecules, while the second one is reduced to water. The basic P450 catalyzed reactions include hydroxylation of sp3-C atoms, heteroatom oxygenation, epoxidation of double bonds, and dealkylation (heteroatom release) (Cryle et al. 2003). Two electrons in form of hydrid ions required for the P450 catalysis are in most cases delivered from the pyridine cofactors NAD(P)H via flavoprotein and/or iron-sulfur redox partners. Traditionally, P450s are classified depending on the topology of the protein components involved in the electron transfer from NAD(P)H to the heme iron. The classical system distinguished two major classes of P450s: Mitochondrial and most bacterial P450s (class I) comprising an iron-sulfur ferredoxin and a FAD-containing ferredoxin reductase; and microsomal P450s (class II) with a membrane bound, FAD- and FMN-containing diflavin reductase. However, the recent discovery of a vast variety of novel P450s has introduced many new redox partner systems. Consequently, a more detailed classifying system is required, for example, a classification comprising ten different P450 classes that was recently suggested by Bernhardt and coworkers (Hannemann et al. 2007).

The monooxygenase described in this paper - CYP109B1 - was identified in 1997, when the complete genome of the *B. subtilis* strain 168 was sequenced (Kunst et al. 1997). The closest relative monooxygenase is CYP109A1 from the *B. subtilis* strain W23, which demonstrates 43% identity to CYP109B1 (Lawson et al. 2004a). CYP109B1 has a potential for the production of secondary metabolites, since its ability to hydroxylate compactin to pravastatin (Endo et al. 2000) and to oxidize testosterone at position C15β (hydroxy-group) and C17 (keto-group) (Agematu et al. 2006; Arisawa and Agematu 2007) has been reported. Furthermore, FT-ICR/MS analysis applied for all P450s originating from *B. subtilis* demonstrated the ability of CYP109B1 to oxidize androsta-1,4-diene-3,17-dione, norethindrone, methyltestosterone, testosterone enanthate, betamethasone dipropionate and medroxyprogesterone acetate. However, the oxidation products for the latter substrates have not been identified (Furuya et al. 2008). Recently we have demonstrated oxidation of (+)-valencene to nootkatol and (+)-nootkatone catalyzed by CYP109B1 (Girhard et al. 2009). (+)-Nootkatone is a high added-value commercial flavoring (Fraatz et al. 2009), which reveals also a potential use of CYP109B1 for the production of fine chemicals in flavor and fragrance industries.

Although there is obviously a significant potential for the application of CYP109B1 in biotechnological processes, a systematic analysis concerning substrate binding and conversion has not been carried out so far. Furthermore, there is also a lack of knowledge regarding the electron transfer partners for this P450. Since the natural electron transfer proteins for CYP109B1 have not been identified yet, "artificial" redox systems with putidaredoxin reductase (PdR) and putidaredoxin (Pdx) from *Pseudomonas putida* were employed in all published studies mentioned above (Agematu et al. 2006; Arisawa and Agematu 2007; Endo et al. 2000; Furuya et al. 2008; Girhard et al. 2009).

Herein we report the identification and application of various electron transfer proteins of CYP109B1 for the significant improvement of substrate conversion in comparison to the

PdR-Pdx system. Furthermore, binding and oxidation of various substrates by CYP109B1 was investigated (Figure 2-1), and oxidized products were identified. The screening of substrates included saturated fatty acids **1-8**, fatty acid methyl- **9-11** and ethyl esters **12-13**, unsaturated fatty acids **14-16**, primary *n*-alcohols **17-19** and *n*-alkanes **20-22**, the sesquiterpene (+)-valencene **23**, the terpenoids α-ionone **25** and β-ionone **26**, the heteroaromatic compounds coumarine **27**, naphthalene **28**, and indole **29**, as well as the steroid testosterone **30**.

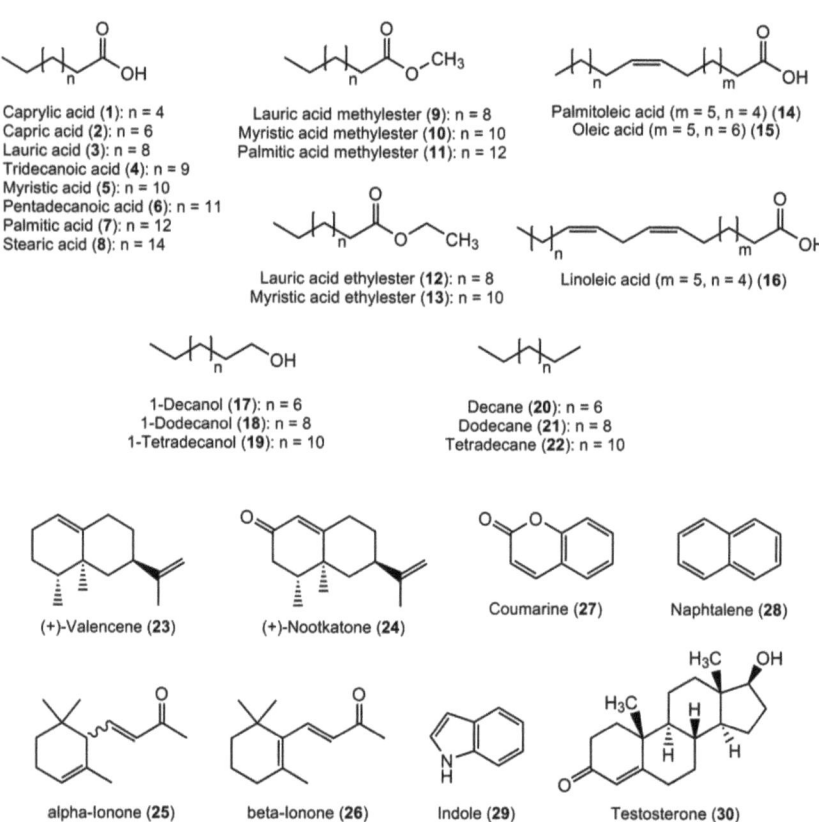

Figure 2-1: Substances investigated with CYP109B1.

2.1.1.3. Materials and methods

Bacterial strains, expression vectors, enzymes and chemicals

E. coli DH5α (F⁻ *supE44 ΔlacU169 (φ80lacZΔM15) hsdR17 recA1 endA1 gyrA96 thi-1 relA1*; Invitrogen, Karlsruhe, Germany) was used as host for cloning purposes. E. coli BL21(DE3) (F⁻ *ompT hsdS$_B$(r_B^- m_B^-) gal dcm* (DE3); Novagen, Darmstadt, Germany) was used for recombinant gene expression. The expression vectors pET11a, pET16b and pET28a(+) were bought from Novagen. Restriction endonucleases, T4 DNA-ligase, *Pfu* DNA-polymerase and Isopropyl-β-D-thiogalactopyranoside (IPTG) were obtained from Fermentas (St. Leon-Rot, Germany). NADH and NADPH were from Codexis (Jülich, Germany). Glucose-6-phosphate dehydrogenase from *Saccharomyces cerevisiae* was from Roche Diagnostics (Mannheim, Germany). Glucose-6-phosphate dehydrogenase from *Leuconostoc mesenteroides* and all other chemicals, solvents and buffer components were purchased from Sigma-Aldrich (Schnelldorf, Germany).

Molecular biological techniques, protein expression and purification

General molecular biology manipulations and microbiological experiments were carried out by standard methods (Sambrook and Russell 2001).

Expression and purification of CYP109B1, PdR and Pdx

Constructions of the following plasmids, protein expressions and purifications were carried out as described previously (Girhard et al. 2009): pET28a-CYP109B1 for expression of the CYP109B1-encoding *yjiB* gene (GeneBank CAB13078) from *B. subtilis* strain 168 DSM 402 (GeneBank NC_000964); pET28a-camA for expression of the PdR-encoding *camA* gene (EMBL-Bank BAA00413) from *P. putida* ATCC 17453; and pET28a-camB for expression of the Pdx-encoding *camB* gene (EMBL-Bank BAA00414) from *P. putida* ATCC 17453.

Expression and purification of the reductase domain of CYP102A1

The plasmid pET28a-CYPRed for expression of the reductase domain (BMR) of CYP102A1 from *B. megaterium* (P450$_{BM3}$) was constructed by insertion of the part of the CYP102A1 gene (GeneBank J04832.1) coding for the diflavin reductase into pET28a(+). Assembly of the construct, protein expression and purification has been described previously (Girhard et al. 2007; Maurer et al. 2003).

Expression and purification of flavodoxin reductase and flavodoxin from E. coli

Flavodoxin (Fdx) from *E. coli* JM109 was expressed from pET11a containing the corresponding gene. Expression and purification was carried out as described elsewhere (Jenkins and Waterman 1998).

Flavodoxin reductase (FdR) from *E. coli* JM109 was inserted into pET16b as follows: The plasmid pET11a-*fpr* as described (Jenkins and Waterman 1998) was utilized as DNA-template for a polymerase chain reaction (PCR), together with the forward 5'-GTATcc**atgg**GC<u>CATCATCATCATCATCAT</u>GCTGATTGGGTA-3', which introduced an *Nco*I site at the start codon and six additional histidines (underlined), and the reverse primer 5'-GATA**ggatcc**TTACCAGTAATGCTCCGCTGTCAT-3', which introduced *Bam*HI at the 3' end. The PCR was performed with *Pfu* DNA-polymerase under the following conditions: 95°C for 3 min, 30 cycles of (95°C for 1 min, 57 ± 2°C for 45 s, 72°C for 2 min), and finally 72°C for 3 min. The PCR-product was purified, digested with endonucleases *Nco*I and *Bam*HI and inserted into previously linearized pET16b. The resulting DNA-construct, pET16b-FdR, was sequenced by automated DNA sequencing (GATC-Biotech, Konstanz, Germany) and encodes for N-terminally His6-tagged FdR. For expression of recombinant FdR, *E. coli* BL21(DE3) cells were transformed with pET16b-FdR, spread onto Luria broth (LB) agar plates containing 100 µg ml^{-1} ampicillin and grown overnight at 37°C. 400 ml of LB-medium supplemented with 100 µg ml^{-1} ampicillin were inoculated with 2 ml of an overnight culture - grown from a single colony - and incubated at 37°C, 180 rpm to an optical density at 600 nm of approximately 0.6. 100 µM IPTG were added and the culture was grown for another 19 h at 25°C, 150 rpm. Cells were harvested by centrifugation at 10800 x g for 20 min at 4°C. The supernatant was discarded and cell pellet was resuspended in 10 ml purification buffer (50 mM Tris-HCl, pH 7.5, 500 mM NaCl, 5% glycerol, 100 µM phenylmethanesulfonyl fluoride (PMSF)). Cells were disrupted by sonication on ice (6 x 30 s bursts, interspaced by 1.5 min), cell debris was removed by centrifugation, and the soluble protein fraction was recovered and filtered through a 0.45 µm filter. FdR was purified from the soluble protein fraction by immobilized metal affinity chromatography (IMAC) using a Talon® resin (7 ml bed volume). The protein lysate was applied to the column, which was pre-equilibrated with 6 column volumes of purification buffer. Non-specifically bound proteins were washed off the column with 4 column volumes of purification buffer containing 20 mM imidazol, before the bound protein was eluted with 2 column volumes of purification buffer with 100 mM imidazol. Fractions containing FdR were pooled, dialyzed twice against 1.5 l of 50 mM Tris-HCl, pH 7.5, 5% glycerol, 100 µM PMSF, 100 µM dithiothreitol at 4°C. Aliquots of the purified protein were prepared and frozen at -20°C until use.

Expression and purification of AdR and Adx

The mammalian truncated Adx_{4-108} (Uhlmann et al. 1994) and AdR were expressed and purified as described before (Hannemann et al. 2002; Sagara et al. 1993).

Expression and purification of YkuN and YkuP

Oligonucleotide primers for the amplification of the *ykuN* and *ykuP* genes from *B. subtilis* strain 168 were designed through analysis of the respective gene sequences using the "SubtiList" genome database (http://genolist.pasteur.fr/SubtiList/). pET16b-YkuN(+) – a vector for the expression of recombinant YkuN – was constructed as follows: The *ykuN*-gene was amplified from genomic DNA of the *B. subtilis* strain 168 by PCR using a forward 5'-GAActcgagATGGCTAAAGCCTTGATTAC-3' and reverse primer 5'-GCCggatcc**TTA**TGAAACATGGATTTTTTCC-3'. PCR conditions were as follows: 95°C for 5 min, 25 cycles of (95°C for 1 min, 57°C for 30 s, 72°C for 2 min), 72°C for 3 min. The PCR-product was purified, digested with endonucleases *Xho*I and *Bam*HI and inserted into previously linearized pET16b. The resulting DNA-construct encodes for N-terminally His10-tagged YkuN under control of the IPTG-inducible T7 phage-promoter.

The basic steps for construction of pET16b-YkuP(+) - a vector for the expression of recombinant YkuP - were identical with those of pET16b-YkuN(+), except that the forward 5'-GAActcgagGCGAAGATTTTGCTCGTTTATG-3' and reverse primer 5'-CCggatcc**CTA**CCTCATTACTGTATCAAAGG-3' were used for PCR.

Both plasmid constructs were verified by automated DNA sequencing. The DNA sequence alignments of the obtained sequences for *ykuN* were in agreement with those reported in the "SubtiList" genome database. Furthermore, the sequencing of *ykuP* confirmed the absence of one adenine nucleotide at position 436 - as discovered by Lawson et al. (Lawson et al. 2004b) - leading to a change in reading frame to obtain a corrected *ykuP* gene. This gene codes for a protein of 151 amino acids with a molecular mass of 16.9 kDa, instead of 178 amino acids and 20.2 kDa as predicted from genome sequencing (Kunst et al. 1997).

Expression of recombinant YkuN and YkuP, IMAC-purification and dialysis were carried out as described for FdR, but with 50 mM Tris-HCl, pH 7.5, 5% glycerol, 100 µM PMSF as dialysis buffer. Aliquots of the purified proteins were stored at -20°C until use.

Determination of protein concentration

The expression level of CYP109B1 was estimated using the CO-difference spectral assay as described previously with $\varepsilon_{450\text{-}490} = 91$ mM^{-1} cm^{-1} (Omura and Sato 1964a; Omura and Sato 1964b). (The ferrous-CO spectrum of CYP109B1 is shown in the supplementary material, Figure 2-6). Using this value, the extinction coefficient for the ferric resting state of CYP109B1 was calculated to be $\varepsilon_{418} = 128$ mM^{-1} cm^{-1}, and was applied for the determination of spin-state shifts upon substrate binding.

The concentration of PdR was determined as the average of the concentration calculated from each of the three wavelength 378, 454 and 480 nm using extinction coefficients (ε) 9.7, 10.0 and 8.5 mM^{-1} cm^{-1} (Purdy et al. 2004). The concentration of Pdx was determined as the average concentration calculated with $\varepsilon_{415} = 11.1$ mM^{-1} cm^{-1} and $\varepsilon_{455} = 10.4$ mM^{-1} cm^{-1} (Purdy et al. 2004). The concentration of BMR was calculated as average concentration determined from $\varepsilon_{452} = 10.0$ mM^{-1} cm^{-1} and $\varepsilon_{466} = 10.0$ mM^{-1} cm^{-1} (Sevrioukova et al. 1996).

Fdx concentration was estimated at 465 nm with $\varepsilon = 8420$ M^{-1} cm^{-1} (Fujii and Huennekens 1974). For FdR the concentration was determined at 456 nm with $\varepsilon = 7.1$ mM^{-1} cm^{-1} (McIver et al. 1998). Adx and AdR concentrations were determined using the molar extinction coefficients $\varepsilon_{415} = 9.8$ mM^{-1} cm^{-1} (Huang and Kimura 1973) and $\varepsilon_{450} = 10.9$ mM^{-1} cm^{-1} (Chu and Kimura 1973), respectively. YkuN and YkuP concentrations were determined at 461 nm with $\varepsilon = 10$ mM^{-1} cm^{-1} (Lawson et al. 2004b; Wang et al. 2007).

Spin-state shift and substrate dissociation constant determinations

Spin-state shifts upon substrate binding were assayed at 25°C under aerobic conditions using an UV-VIS scanning photometer (UV-2101PC, Shimadzu, Japan) equipped with two tandem quartz cuvettes (Hellma, Müllheim, Germany) simultaneously. One chamber of each cuvette contained 10 µM CYP109B1 in buffer (50 mM TrisHCl, pH 7.5), whereas the second chamber contained buffer alone. The initial volume was 800 µl. Substrates were dissolved in DMSO, except for stearic acid **8** and testosterone **31**, which were dissolved in ethanol. Substrate titrations where done by adding small (< 5 µl) aliquots of an appropriate stock of a substrate or ligand into the P450 containing chamber of the sample cuvette. An equal amount of solvent was added into the buffer containing chamber of the reference cuvette and spectral changes between 350 and 500 nm were recorded. When saturation of CYP109B1 with a substrate was achieved, substrate dissociation constants (K_D) were calculated by fitting the data to the hyperbolic equation (Eqn. 1):

$$\Delta A = \frac{\Delta A_{max} \times [S]}{K_D + [S]} \qquad \text{Eqn. 1}$$

ΔA is the peak-to-trough difference in absorbance, ΔA_{max} is the maximum difference in absorbance and [S] is the concentration of the substrate.

Reconstitution of CYP109B1 activity in vitro

Various enzyme systems were applied to reconstitute the *in vitro* activity of CYP109B1. The basic setup for determination of the NAD(P)H consumption rate and coupling of the reconstituted systems was as follows: 1 µM reductase, 10 µM flavo- or ferredoxin, 1 µM CYP109B1, 4 mM glucose-6-phosphate ,1 mM $MgCl_2$, 200 µM substrate (from a 10 mM stock solution in DMSO or ethanol) were mixed in 50 mM TrisHCl, pH 7.5 to a final volume of 250 µl in a 96 microwell plate. After incubation at 30°C for 2 min, 200 µM NAD(P)H were added and the absorption was followed at 340 nm. The NAD(P)H-consumption rate was calculated using $\varepsilon = 6.22\ mM^{-1}\ cm^{-1}$. After all NAD(P)H was consumed, the internal standard (final concentration 50 µM) was added and samples were prepared for quantitative GC/MS analysis to determine the coupling efficiency.

For product identification and determination of conversion the basic setup was as follows: 1 µM reductase, 10 µM flavo- or ferredoxin, 1 µM CYP109B1, 5 units of glucose-6-phosphate dehydrogenase from *S. cerevisiae* (for regeneration of NADPH) or from *L. mesenteroides* (for regeneration of NADH), 4 mM glucose-6-phosphate and 1 mM $MgCl_2$ were mixed in a 1.5 ml reaction tube and incubated at 30°C for 2 min. 50 mM TrisHCl, pH 7.5 and 200 µM substrate were added to a final volume of 500 µl. The reaction was started by addition of 200 µM NADH or NADPH and samples were incubated at 30°C for 20 min - 2 h. After incubation the internal standard (final concentration 50 µM) was added and samples were prepared for product analysis.

For saturated -, unsaturated fatty acids and primary *n*-alcohols the reaction was stopped by addition of 20 µl 37% HCl. Internal standards were used as follows: **4** in the case of **5, 6, 7, 8, 10, 11** and **13**; **2** in the case of **3, 4, 9** and **12**; **1** in the case of **2**; **17** in the case of **18** and **19**; **19** in the case of **17**. The reaction mixtures were extracted with 1 ml diethyl ether, dried over anhydrous $MgSO_4$ and the organic phases were evaporated. The residues were dissolved in 40 µl N,O-bis(trimethylsilyl)trifluoroacetamide containing 1% trimethylchlorosilane for derivatization, transferred into GC vials and incubated at 80°C for 30 min prior to analysis.

Samples containing compounds **20-30** were extracted with 400 µl ethyl acetate. (-)-Carvone was used as internal standard for **23**, **25** served as internal standard for **26** and vice versa. The organic phases were recovered and transferred into GC vials for product analysis.

Product analysis

The analysis of conversion products was carried out by gas-liquid chromatography mass spectrometry (GC/MS) on a GC/MS-QP2010 (Shimadzu, Tokyo, Japan), equipped with a FS-Supreme-5 column (30 m x 0.25 mm x 0.25 µm, Chromatographie Service GmbH, Langerwehe, Germany). The injector and detector temperatures were set at 250°C and 285°C, respectively, with helium as carrier gas. 1 µl of a sample was injected for analysis. Conversion products were identified by their characteristic mass fragmentation patterns.

For quantification of the substrates the detector response was calibrated with internal standards. Therefore, mixtures of 50 mM TrisHCl buffer, pH 7.5 containing the respective substrate for calibration in final concentrations of 10 to 200 µM together with the respective internal standard in a final concentration of 50 µM were treated as described for normal conversion reactions and analyzed by GC/MS. The ratio of the area of the substrate to that of the internal standard was plotted against the substrate concentration to give a straight-line calibration plot.

For fatty acids and primary *n*-alcohols a split of 30 was used. For **1, 2, 3, 4, 9, 12, 18** and **19** the column temperature was maintained at 130°C for 2 min, ramped to 250°C at a rate of 10°C min^{-1} and held at 250°C for 3 min. For **5, 6, 7, 8, 10, 11, 13**, and **14-16** the column temperature was maintained at 180°C for 2 min, increased to 300°C at a rate of 8°C min^{-1} and held at 300°C for 5 min. For **17** and **20-22** a starting temperature of 110°C was maintained for 2 min, raised to 220°C at a rate of 10°C min^{-1} and kept at this temperature for 2 min.

23 was injected at a split of 4. The column temperature was maintained at 120°C for 4 min, ramped to 250°C at a rate of 10°C min^{-1} and held at 250°C for 5 min. For **25** and **26** the split was 5. The column temperature was controlled at 130°C for 2 min, then raised to 230°C at a rate of 20°C min^{-1} and kept for 2 min.

In the case of **27** and **28** injections were done with a split of 4. The column temperature was maintained at 120°C for 4 min, increased to 250°C at a rate of 10°C min^{-1} and held at 250°C for 5 min. **29** was injected at a split of 10. The column temperature was maintained at 80°C for 4 min, ramped to 250°C at a rate of 8°C min^{-1} and held for 4 min.

2.1.1.4. Results and discussion

Expression, purification and spectral characterization of CYP109B1

The recombinant expression of CYP109B1 was performed in *E. coli* BL21(DE3) using the expression vector pET28a(+). The monooxygenase was purified by immobilized metal affinity chromatography (supplementary material, Figure 2-7). The expression yield of CYP109B1 determined from CO-difference spectra was 3570 nmol l^{-1} (160 mg l^{-1}). CYP109B1 showed a characteristic absorption at 452 nm for the Fe^{II}(CO) complex with no evidence of the inactive P420 form. Photometric characterization of the purified CYP109B1 revealed spectra typical for a P450 monooxygenase (supplementary material, Figure 2-6).

Determination of substrate binding constants

Substrate binding by P450s is often characterized by a spin-state shift of the Soret band from approximately 418 nm (representing the low-spin substrate free form) to around 390 nm for the high-spin form when a substrate is bound in close proximity to the heme, which results in a classical Type I spectrum (Schenkman et al. 1981). Therefore, this method was used to identify potential CYP109B1 substrates.

A variety of saturated and unsaturated fatty acids showed Type I binding to CYP109B1 (Figure 2-2). The substrate dissociation constants (K_D) were determined as described in the material and methods section. The lowest K_D for saturated fatty acids of 230 µM was calculated for palmitic acid **7**. Saturated fatty acids with shorter or longer chain lengths showed weaker binding and subsequently higher K_D values were calculated for those substrates (Table 2-1). The unsaturated fatty acids **14-16** tested showed in general better binding to CYP109B1 than saturated fatty acids, which is reflected in lower K_D values calculated for those substrates.

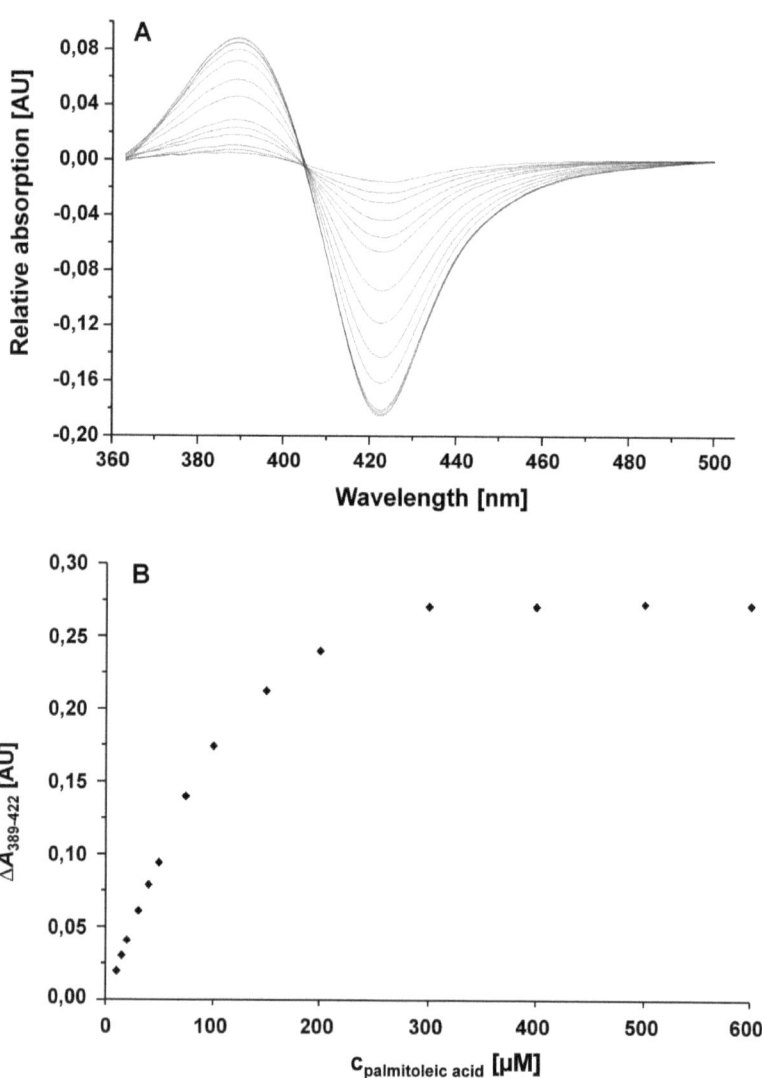

Figure 2-2: (**A**) Type I substrate binding of palmitoleic acid **14** to CYP109B1. Peak and trough were observed at 389 and 422 nm for this substrate, respectively. (**B**) Peak-to-trough absorbance differences plotted against the respective palmitoleic acid concentration assayed.

Type I binding was also observed for α-ionone **25** and β-ionone **26** and K_D-values of 205 or 180 µM were calculated, respectively, indicating that CYP109B1 shows a slight preference for the binding of β-ionone **26** over α-ionone **25**. Tight binding with lowest K_D value of 24.5 µM was calculated for indole **29**, whereas no spin-state shifts were induced for coumarine **27** and naphthalene **28**.

Table 2-1: Type I spin-state shift values and binding constants for CYP109B1.

Type of substrate	Name (**Number**)	K_D [µM]	Spin-state shift [%][a]
Saturated fatty acids	caprylic acid (**1**)	(+)	(+)
	capric acid (**2**)	(+)	(+)
	lauric acid (**3**)	3670 ± 440	25
	tridecanoic acid (**4**)	1970 ± 250	35
	myristic acid (**5**)	1190 ± 100	55
	pentadecanoic acid (**6**)	720 ± 55	15
	palmitic acid (**7**)	228 ± 27	10
	stearic acid (**8**)	983 ± 67	7
Unsaturated fatty acids	palmitoleic acid (**14**)	115 ± 13	8
	oleic acid (**15**)	50 ± 3	20
	linoleic acid (**16**)	12 ± 1	25
Terpenoids	(+)-valencene (**23**)	—	—
	(+)-nootkatone (**24**)	320 ± 11	10
	α-ionone (**25**)	205 ± 6	25
	β-ionone (**26**)	180 ± 16	20
Hetero-aromatic	coumarine (**27**)	—	—
	naphthalene (**28**)	—	—
	indole (**29**)	25 ± 3	< 5
Steroids	testosterone (**30**)	—	—

[a] Estimated to within ≤ 5%
— No spin-state shift observed
(+) A spin-state shift was observed for these substances, but a binding constant could not be calculated, since substance solubility reached its limit before complete saturation of CYP109B1 was achieved

As seen in Table 2-1 none of the substrates tested induced a full spin-state shift from low-spin in the absence of substrate to high-spin. The strongest shift of ~ 55% was observed for myristic acid. Other substrates resulted in lower high-spin heme content. In most cases, the percentage of the enzyme in high-spin state correlate to the calculated binding constants.

Interestingly, no spin-state shifts were observed when testosterone 30 or (+)-valencene 23 were used for the binding experiment, although these substances are converted by CYP109B1 as reported previously (Agematu et al. 2006; Girhard et al. 2009). Furthermore, when the product of the (+)-valencene 23 conversion - (+)-nootkatone 24 - was added, classical Type I binding could be observed and a K_D of 320 µM was calculated (Table 2-1). This phenomenon has also been reported for CYP102A1 from *B. megaterium*, which is also capable of converting (+)-valencene 23 (Sowden et al. 2005).

Reconstitution of CYP109B1 activity in vitro

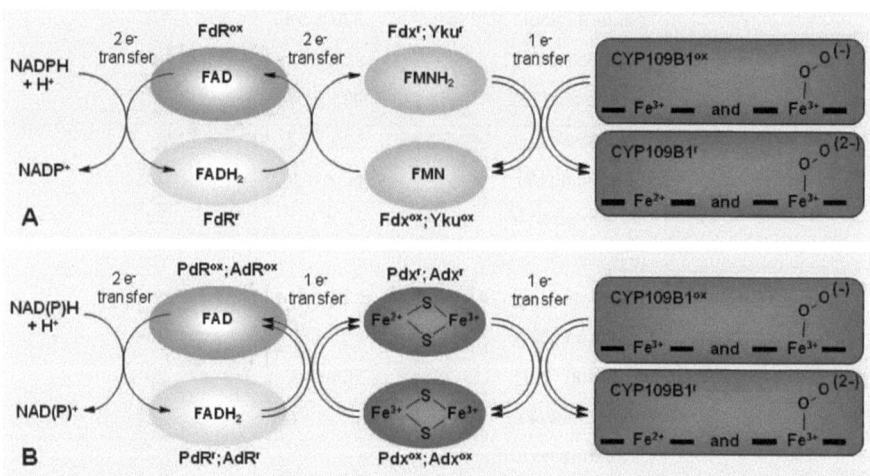

Figure 2-3: Schematic organization of different redox systems for CYP109B1 applied in this study and their classification according to (Hannemann et al. 2007). (**A**) FdR-Fdx, FdR-YkuN and FdR-YkuP (class III, bacterial systems); (**B**) PdR-Pdx (class I, bacterial system) and AdR-Adx(4-108) (class I, mitochondrial system). The BMR system was not included, since it did not support activity of CYP109B1. ^{ox}, oxidized form of the respective enzyme; ^{r}, reduced form.

CYP109B1 requires - like most P450s in general - electron transfer partners for activity. Since natural partners for CYP109B1 have not been identified yet, *in vitro* activity of CYP109B1 was reconstituted with various reductases and ferredoxins or flavodoxins that have been described to support activity of other P450s previously (Figure 2-3).

It has been shown that FdR-Fdx from *E. coli* supports the activity of P450c17 (Jenkins and Waterman 1998); bovine AdR in combination with the corresponding truncated Adx_{4-108} was used for reconstituting the steroid hydroxylase activity of bacterial CYP106A2 (Virus and Bernhardt 2008; Virus et al. 2006), and the separately expressed reductase domain of CYP102A1 (BMR) was shown to support fatty acid hydroxylating activity of CYP152A2 (Girhard et al. 2007). Furthermore, the flavodoxins YkuN and YkuP from *B. subtilis* have been reported to support activity of P450 BioI in reconstituted systems with FdR from *E. coli* (Lawson et al. 2004b). Thus, FdR-YkuN and FdR-YkuP were also tested for their ability to support activity of CYP109B1. The NAD(P)H consumption rates, coupling efficiencies and conversions of myristic acid **5** were determined with these systems and compared to the PdR-Pdx system used previously for activity reconstitution of CYP109B1 (Table 2-2).

Table 2-2: NAD(P)H consumption, coupling and conversion of myristic acid **5** by CYP109B1 in reconstituted enzyme systems with a ratio of reductase : ferredoxin(flavodoxin) : CYP109B1 of 1:10:1.

Reconstituted system	NAD(P)H consumption rate[a]	Coupling [%]	Conversion [%][b]
PdR-Pdx	6.6 ± 0.3	2.4 ± 0.3	16.0 ± 1.4
BMR	7.3 ± 0.4	0.0	0.0
AdR-Adx	20.2 ± 1.2	45.8 ± 3.5	99.1 ± 0.7
FdR-Fdx	3.0 ± 0.1	7.5 ± 0.9	52.3 ± 4.0
FdR-YkuP	20.3 ± 1.0	1.8 ± 0.1	13.9 ± 0.9
FdR-YkuN	19.3 ± 0.2	12.3 ± 0.5	90.2 ± 6.6

[a] Rates are given in nmol (nmol CYP109B1)$^{-1}$ min^{-1}; the background consumption rate with CYP109B1 alone was 0.6 ± 0.1
[b] Conversion was determined after 2 h with NAD(P)H regeneration

For the PdR-Pdx system 16% conversion of myristic acid **5** was observed within 2 h at 30°C using a ratio of PdR:Pdx:CYP109B1 of 1:10:1. In comparison, conversion of myristic acid **5** was 52% with FdR-Fdx and 99% with AdR-Adx under the same conditions. For FdR-YkuN 90% of myristic acid **5** were converted, whereas for FdR-YkuP only 14%

conversion was achieved, respectively. Systems containing BMR or FdR alone - without YkuN or YkuP - did not show any activity against myristic acid **5** (Table 2-2).

The overall efficiency of a reconstituted system depends *inter alia* on its NAD(P)H consumption rate and its coupling efficiency. Since two independent proteins were used for electron supply to CYP109B1, a loss of electrons can occur either between the reductase and the ferre-/flavodoxin and/or between the ferre-/flavodoxin and CYP109B1. In both cases, it results in uncoupling between NAD(P)H consumption and product formation. Similar NADPH consumption rates of ~ 20 nmol (nmol CYP)$^{-1}$ min^{-1} were seen for AdR-Adx, FdR-YkuP and FdR-YkuN, the coupling efficiencies of these systems, however, differed strongly from each other (Table 2-2). The highest coupling of 45.8% was observed for AdR-Adx, whereas coupling for FdR-YkuP was 12.3%, and for FdR-YkuP 1.8% only. If one compares the conversion values achieved for myristic acid **5** under regeneration of NAD(P)H, where the cofactor was not rate limiting, it is obvious that the coupling efficiency of a reconstituted system is of greater importance for overall performance than the corresponding NAD(P)H consumption rate.

We also determined the conversion of (+)-valencene **23** with the reconstituted systems. In general, lower conversions than of myristic acid **5** were reached, but the tendency was similar. For (+)-valencene **23**, 10% conversion was observed with PdR-Pdx and in comparison, 17% and 23% with FdR-Fdx and AdR-Adx, respectively. For FdR-YkuN and FdR-YkuP conversions were 26% and < 1% (data not shown). The lower coupling efficiencies of the latter two systems explain the lower conversion of myristic acid **5** and (+)-valencene **23**.

From the reconstituted systems tested, AdR-Adx and FdR-YkuN are more efficient electron transfer partners than PdR-Pdx, FdR-Fdx and FdR-YkuP, as they demonstrate higher NAD(P)H consumption rates, higher coupling and hence significantly higher conversions were achieved.

The flavodoxin YkuN can be considered as a potential natural redox partner for CYP109B1 in *B. subtilis*, however, in the absence of the appropriate reductase from the same strain this hypothesis cannot be proved. Nevertheless, since the relative ratios of P450 to redox partner play an important role for the overall rate of substrate oxidation, an empirical approach with FdR-YkuN was undertaken in order to improve the substrate oxidation rates while conserving enzyme. As the rate-limiting step in P450-mediated oxidation is often the transfer of the second electron to an oxy-P450-substrate complex, we aimed to maximize the concentration of reduced YkuN by varying its concentration and that of FdR. When conversion of myristic acid **5** and (+)-valencene **23** was investigated with these systems it was observed that the YkuN concentration represented a bottleneck for the *in vitro* activity reconstitution of CYP109B1. For example, the conversion of (+)-valencene **23** increased from 10% up to ~ 30%, when the ratio was changed from 1:1:1 to 1:20:1 (Table 2-3). Further increase of the flavodoxin concentration up to a ratio of 1:40:1 resulted, however, in only slightly higher conversion of ~ 31%, which means that electron transfer by YkuN was not rate limiting any longer. It was also observed that additional FdR did not show any significant increase in the conversion of (+)-valencene **23**. A similar tendency was observed during oxidation of myristic acid **5**. However, taking into account that the conversions achieved with FdR-YkuN were still lower than those achieved with AdR-Adx, and that high amounts of YkuN were needed, the AdR-Adx system was used for further experiments.

Table 2-3: Conversion of myristic acid **5** and (+)-valencene **23** by CYP109B1 in reconstituted FdR-YkuN systems with varying ratios of redox partner proteins.

FdR:YkuN:CYP [µM]	Conversion [%][a]	
	Myristic acid **5**	(+)-Valencene **23**
1:0:1	—	—
1:1:1	38.5	10.0
1:2:1	70.2	10.8
1:5:1	88.7	19.3
1:10:1	90.2	25.5
1:20:1	nd	30.1
1:40:1	nd	31.2
2:5:1	86.3	20.6
2:10:1	94.0	26.2
2:20:1	nd	29.0
2:40:1	nd	34.2
10:10:1	93.5	26.4

[a] Conversion was determined after 2 h with NAD(P)H regeneration; values represent the mean of three experiments, with standard deviations within ≤ 10% of the mean
— No conversion; *nd* not determined

Conversion of fatty acids

The ability to oxidize fatty acids is well known for a variety of bacterial P450s. Among well-characterized fatty acid hydroxylases are the natural fusion enzymes from the CYP102A-family, namely CYP102A1 (P450$_{BM-3}$) from *B. megaterium* (Miura and Fulco 1975) and its two homologues from *B. subtilis*, CYP102A2 (Budde et al. 2004) and CYP102A3 (Gustafsson et al. 2004).

CYP109B1 was found to oxidize saturated fatty acids with a chain length ranging from C10 to C18. Highest conversions of more than 98% were achieved for tridecanoic- **4** (C13), myristic- **5** (C14) and pentadecanoic acid **6** (C15), whereas conversion decreased for the substrates of shorter or longer chain length (supplementary material, Table 2-5). Interestingly, there was no obvious correlation between the measured substrate dissociation constants and the achieved conversions. Palmitic acid **7** (C16), for example, for which the lowest K_D value was determined, was oxidized by only 51%.

The products of fatty acid oxidations were identified by their characteristic mass fragmentation patterns after derivatization with trimethylchlorosilane (supplementary material, Figure 2-8). The GC/MS analysis demonstrated that CYP109B1 hydroxylates fatty acids at subterminal positions yielding mono-hydroxylated products exclusively. Quantitative analysis of these products revealed that the carbon atoms C11 and C12 counted from the carboxyl group of all fatty acids were preferred for hydroxylation independently on chain length.

Further we checked activity of CYP109B1 towards methyl- **9-11** or ethyl esters **12,13** of fatty acids. Surprisingly, the monooxygenase was active towards these substrates, but with lower activity (supplementary material, Table 2-5). For example, 95% of lauric acid **3**, but only 78% of the corresponding methyl ester **9** and 32% of the ethyl ester **12** were converted. A similar tendency (99% versus 80% and 34%, respectively) was observed for myristic acid **5** and its esters **10,13**. The regioselectivities of the hydroxylation observed for the esterified fatty acids was identical to those of the respective non-esterified fatty acid.

All unsaturated fatty acids tested (**14-16**) were oxidized by CYP109B1 with high activities (> 95% within 15 min). Qualitative GC/MS analysis revealed that the oxidation proceeded unselectively resulting in more than 20 products for each substrate, which mostly could not be identified (and therefore are not included).

Conversion of n-alkanes and primary n-alcohols

Hydroxylation of alkanes has been described for a variety of P450s and generated mutants thereof (Bernhardt 2006 and references cited therein). Furthermore, primary n-alcohols are reported as P450 substrates as well, for example 1-dodecanol **18** is hydroxylated at seven different positions by P450s from the white-rot fungus *Phanerochaete chrysosporium* (Matsuzaki and Wariishi 2004) and 1-tetradecanol **19** is oxidized by CYP102A1 at position ω_{-1} to ω_{-3}, with preference for the ω_{-2} carbon atom (Miura and Fulco 1975). Therefore, the *n*-alkanes decane **20** (C10), dodecane **21** (C12) and tetradecane **22** (C14) and their 1-hydroxylated derivatives 1-decanol **17**, 1-dodecanol **18** and 1-tetradecanol **19** were investigated as potential substrates for CYP109B1.

CYP109B1 did not show any activity against the tested *n*-alkanes **20-22**, but hydroxylated 1-dodecanol **18** and 1-tetradecanol **19** at six different subterminal carbon atoms (ω_{-1} to ω_{-6}), with similar regioselectivity for both substrates (Table 2-4). Conversion of 1-tetradecanol **19** was slightly lower as compared with 1-dodecanol **18** (12% and 17%, respectively). 1-Decanol **17** was oxidized with high activity (99.5% conversion) at subterminal ω_{-1} to ω_{-5} positions. The terminal carbon atom of any of the alcohols was not oxidized and - in contrast to fatty acids - there was always a preference for hydroxylation at the ω_{-2} carbon atom, independent from the chain length (Table 2-4). Furthermore, the regioselectivities of hydroxylation of 1-decanol **17** and capric acid **2** were completely different, since capric acid **2** was oxidized at ω_{-1} position exclusively.

Table 2-4: Product distribution and conversion of primary *n*-alcohols **17-19** by CYP109B1.

Name (Number)	Conversion [%]	Regioselectivity [%][a]					
		ω_{-6}	ω_{-5}	ω_{-4}	ω_{-3}	ω_{-2}	ω_{-1}
1-decanol (**17**)	99.5 ± 0.2	—	2.9	7.2	22.2	40.5	27.3
1-dodecanol (**18**)	16.6 ± 3.1	4.4	4.9	17.4	18.7	27.2	27.5
1-tetradecanol (**19**)	11.8 ± 0.3	5.9	6.7	18.8	17.1	26.8	24.9

[a] Values represent % of the total product and are the mean of three experiments, with standard deviations within ≤ 5% of the mean
— Compound was not observed or below the detection limits (1 µmol l^{-1})

Conversions for 1-dodecanol **18** and 1-tetradecanol **19** were much lower than those for the corresponding fatty acids, lauric- **3** and myristic acid **5**, and also lower than those of the corresponding fatty acid esters **9,10,12,13**, whereas conversion of 1-decanol **17** was significantly higher than those of capric acid **2**.

Conversion of ionones

Ionones are valuable fragrance constituents and therefore have attracted attention of flavor and fragrance industries, like manufacturers of perfumes, cosmetics and other fine chemicals. Their oxygenated derivatives are sought-after building blocks, for example 4-hydroxy-β-ionone **32**, which is an important intermediate for the synthesis of carotenoids or the plant hormone abscisic acid (Meyer 2002). CYP109B1 was shown to oxidize the sesquiterpenoid analogs α-ionone **25** and β-ionone **26** with high activity. Using AdR-Adx, the conversion achieved 70% (± 6%) for α-ionone **25** and 91% (± 2%) for β-ionone **26**, within 2 h.

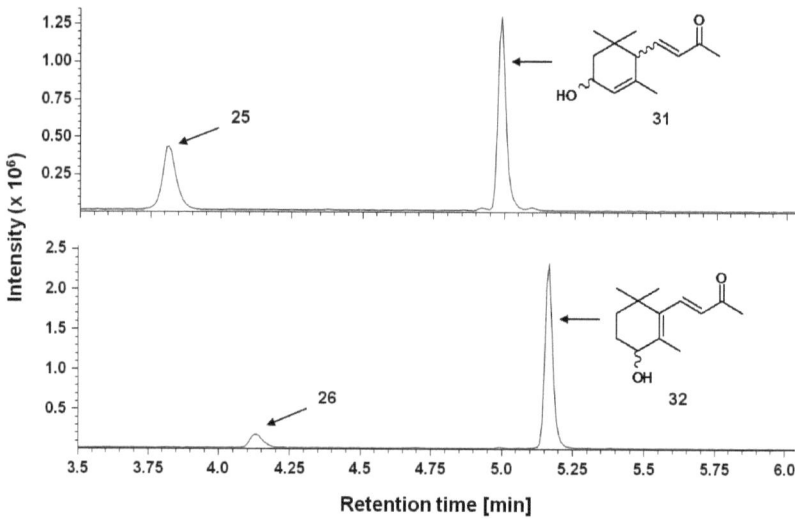

Figure 2-4: GC-diagram of α-ionone **25** and β-ionone **26** conversions by CYP109B1. The reaction products were identified as 3-hydroxy-α-ionone **31** and 4-hydroxy-β-ionone **32**.

Furthermore, a single product peak was observed by GC/MS analysis on an achiral column for each of the substrates, correspondingly (Figure 2-4). For β-ionone **26** the oxidation product was identified as 4-hydroxy-β-ionone **32** by comparison of the MS with those of an authentic reference substance reported previously (Urlacher et al. 2006). For α-ionone **25**, the oxidation product was identified as 3-hydroxy-α-ionone **31** in the same way by comparison with MS reported elsewhere (Celik et al. 2005). Since conversion of racemic α-ionone **25** was higher than 50%, obviously both enantiomers were converted by CYP109B1. The unselective hydroxylation of racemic α-ionone **25** should result in a mixture of four diastereoisomers, and consequently two isomers would be detected by achiral analysis (Celik et al. 2005). The observed single peak might indicate that the enzyme is enantioselective. However, this aspect requires further detailed investigation.

Indole, coumarine and naphthalene

CYP109B1 did not show any activity towards coumarine **27** and naphthalene **28** regardless of the redox-system applied. Indole induced a change of the heme spin state from low-spin in the absence of substrate to only 3% high-spin. Consequently, no activity towards indole **29** was observed, although this compound shows a high affinity to CYP109B1 confirmed by the low K_D value of 24.5 µM.

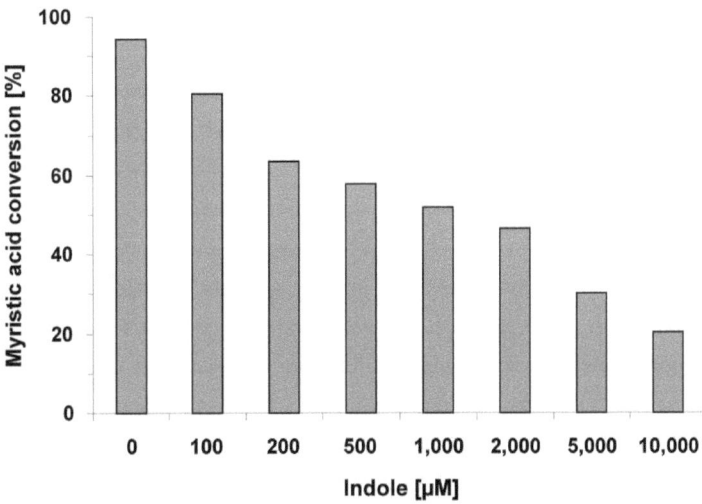

Figure 2-5: Inhibition of myristic acid **5** oxidation by CYP109B1 through indole **29**. Conversion was carried out for 1 h at 30°C employing the AdR-Adx system. The data represent the average values of two experiments carried out in triplicates.

Since indole **29** binds to CYP109B1, but is not oxidized, we investigated, if it could serve as competitive inhibitor for CYP109B1. Conversion of myristic acid **5** - whose K_D is approximately 50-fold higher than those of indole **29** - was monitored under addition of indole **29**. As expected, the presence of indole **29** in the reaction mixture drastically reduced the conversion of myristic acid **5** by CYP109B1 (Figure 2-5). For example, when equal amounts (200 µM) of myristic acid **5** and indole **29** were present in the reaction mixture, the conversion dropped from ~ 94% (negative control without indole **29**) to ~ 64%. Higher amounts of indole **29** reduced conversion even further; only 20% of myristic acid **5** were converted in the presence of 10 mM indole **29**. Indole can thus be referred to as a competitive inhibitor for CYP109B1. Binding of an inhibitor to a P450 enzyme would normally result in a Type II binding spectrum, with a Soret maximum at 425-435 nm and a trough at 390-405 nm (Correia and Oritz de Montellano 2005), whereas binding of indole **29** by CYP109B1 gave clearly Type I binding spectra (data not shown).

2.1.1.5. Discussion

With increasing availability of genome sequences and sophisticated heterologous expression methods, new P450s identified by bioinformatics methods are readily expressed and isolated. These new enzymes will greatly expand the range of compounds that can be oxidized, but equally important are the different product selectivities of existing enzymes. Therefore, for the identification of biotechnologically relevant enzymes, genome mining should be supported by substrate profiling and functional characterization of candidate P450s. Since many P450 genes are found without their associated electron transfer proteins immediately upstream or downstream, substrate spectra of such P450s cannot be identified. In some cases the search among annotated possible candidates from the same genome yield effective electron transfer systems for P450s (Ewen et al. 2009; Mandai et al. 2009). If no potential "natural" candidates are found within the genome, however, known electron transfer proteins for other P450s can be applied, because they often demonstrate "cross" interaction with foreign P450s and support their catalysis. The ability to accept electrons from "artificial" flavodoxins and ferredoxins, and the display of "cross" activities has been reported for several P450 monooxygenases (Bell et al. 2009; Bell and Wong 2007; Goni et al. 2009).

Our results demonstrate that the activity of CYP109B1 can be supported by a FAD-containing reductase and either by a FMN-containing flavodoxin or a [2Fe-2S]-containing ferredoxin. The best results for CYP109B1 were observed with bovine AdR and truncated

Adx$_{4-108}$, while activity with putidaredoxin and putidaredoxin reductase from *P. putida* was quite low. Both electron transfer systems belong to the class I of the P450 systems. CYP109B1 demonstrated also activity in combination with one of the two FMN-containing flavodoxins from *B. subtilis* (YkuN), but does not interact with the diflavin reductase of CYP102A1. YkuN can represent a natural electron transfer protein for CYP109B1, however, in combination with the "artificial" reductase from *E. coli*, the overall activity of the reconstituted system was lower compared to AdR-Adx. Structural and thermodynamic characterization of CYP109B1 combined with YkuN should provide insights into the flavodoxin-P450 interaction and are topic of further investigations. As possible candidate the 4Fe-4S ferredoxin from *Bacillus subtilis* will be also tested (Green et al. 2003).

Supported by "artificial" electron transfer partners, CYP109B1 catalyzes oxidation of a broad range of chemically different substrates. Unsaturated fatty acids were oxidized with high activity but low regioselectivity leading to a mixture of numerous products. Upon oxidation of saturated fatty acids and esterified saturated fatty acids, the carbon atoms C11 and C12 counted from the carboxyl group of all substrates were always preferred for hydroxylation, independently on chain length. Moreover, the same regioselectivity of CYP109B1 was observed during hydroxylation of esterified fatty acids. This might indicate that the carboxyl group of fatty acids is not as crucial for substrate recognition by CYP109B1.

Interestingly, CYP109B1 hydroxylates primary *n*-alcohols of the corresponding chain length with completely different regioselectivities, preferring ω$_{-2}$ position. A possible explanation could be that CYP109B1 possesses alternative binding sites or orientations for primary *n*-alcohols and fatty acids.

In contrast to the acyclic substrates, the cyclic sesquiterpenoid analogs α-ionone and β-ionone were hydroxylated at secondary allylic position with 100% regioselectivity, which in the case of β-ionone to our best knowledge has been described only for the mutated P450$_{BM-3}$ A74E F87V P368S (Urlacher et al. 2006). The reason for the high regioselectivity of CYP109B1 for oxidation of ionones might be caused by electronic activation of the secondary allylic carbon atoms. In the case of α-ionone, the carbon atom C3 next to the double bond between C4=C5 is most susceptible to an oxidative attack by CYP109B1, while for β-ionone this holds true for C4 next to the C5=C6 double bond. A similar preference for allylic oxidation has been described previously for the oxidation of (+)-valencene by CYP109B1, which was oxidized with high regioselectivity at allylic C2 position (Girhard et al. 2009). Thus, CYP109B1 seems to be an attractive candidate for selective oxidation of terpenes, generating valuable compounds for flavor and fragrance industries. Especially the combination of CYP109B1 with the AdR-Adx redox system could be valuable for biotechnological applications in a whole-cell bioconversion as shown before for steroid hydroxylation using CYP106A2 together with this redox system (Hannemann et al. 2006).

2.1.1.6. Acknowledgements

We wish to thank Kyoko Momoi, Sumire Honda Malca and Svetlana Tihovsky (Universitaet Stuttgart) for their help in preparation of Fdx and cloning of FdR, YkuN and YkuP. We are also thankful to Wolfgang Reinle for expression and purification of Adx and AdR. MG, TK and VBU acknowledge the support of this work by Deutsche Forschungsgemeinschaft (SFB706) and Ministerium für Wissenschaft, Forschung und Kunst Baden-Württemberg.

2.1.1.7. Conflict of interest

The authors declare that they have no conflict of interest.

2.1.1.8. References (to chapter 2.1.1)

Agematu H, Matsumoto N, Fujii Y, Kabumoto H, Doi S, Machida K, Ishikawa J, Arisawa A (2006) Hydroxylation of testosterone by bacterial cytochromes P450 using the Escherichia coli expression system. Biosci Biotechnol Biochem 70:307-311.

Arisawa A, Agematu H (2007) A Modular Approach to Biotransformation Using Microbial Cytochrome P450 Monooxygenases. In: R. D. Schmid and V. B. Urlacher (eds) Modern Biooxidation, First edn. Wiley-VCH, Weinheim, pp 177-192.

Bell SG, Dale A, Rees NH, Wong LL (2009) A cytochrome P450 class I electron transfer system from *Novosphingobium aromaticivorans*. Appl Microbiol Biotechnol. doi:10.1007/s00253-009-2234-y.

Bell SG, Wong LL (2007) P450 enzymes from the bacterium *Novosphingobium aromaticivorans*. Biochem Biophys Res Commun 360:666-672.

Bernhardt R (2006) Cytochromes P450 as versatile biocatalysts. J Biotechnol 124:128-145.

Budde M, Maurer SC, Schmid RD, Urlacher VB (2004) Cloning, expression and characterisation of CYP102A2, a self-sufficient P450 monooxygenase from Bacillus subtilis. Appl Microbiol Biotechnol 66:180-186.

Celik A, Flitsch SL, Turner NJ (2005) Efficient terpene hydroxylation catalysts based upon P450 enzymes derived from actinomycetes. Org Biomol Chem 3:2930-2934.

Chu JW, Kimura T (1973) Studies on adrenal steroid hydroxylases. Molecular and catalytic properties of adrenodoxin reductase (a flavoprotein). J Biol Chem 248:2089-2094.

Correia MA, Oritz de Montellano PR (2005) Inhibition of Cytochrome P450 Enzymes. In: P. R. Oritz de Montellano (ed) Cytochrome P450 : Structure, Mechanism, and Biochemistry, Third edn. Springer US, Berlin, pp 247-322.

Cryle MJ, Stok JE, De Voss JJ (2003) Reactions catalyzed by bacterial cytochromes P450. Aust J Chem 56:749-762.

Endo H, Yonetani Y, Mizoguchi H, Hashimoto S, Ozaki A (2000) Process for producing HMG-CoA reductase inhibitor. WO/2000/044886.

Ewen KM, Hannemann F, Khatri Y, Perlova O, Kappl R, Krug D, Huttermann J, Muller R, Bernhardt R (2009) Genome mining in *Sorangium cellulosum* So ce56: identification and characterization of the homologous electron transfer proteins of a myxobacterial cytochrome P450. J Biol Chem 284:28590-28598.

Fischer M, Knoll M, Sirim D, Wagner F, Funke S, Pleiss J (2007) The Cytochrome P450 Engineering Database: a navigation and prediction tool for the cytochrome P450 protein family. Bioinformatics 23:2015-2017.

Fraatz MA, Berger RG, Zorn H (2009) Nootkatone--a biotechnological challenge. Appl Microbiol Biotechnol 83:35-41.

Fujii K, Huennekens FM (1974) Activation of methionine synthetase by a reduced triphosphopyridine nucleotide-dependent flavoprotein system. J Biol Chem 249:6745-6753.

Furuya T, Nishi T, Shibata D, Suzuki H, Ohta D, Kino K (2008) Characterization of orphan monooxygenases by rapid substrate screening using FT-ICR mass spectrometry. Chem Biol 15:563-572.

Girhard M, Machida K, Itoh M, Schmid RD, Arisawa A, Urlacher VB (2009) Regioselective biooxidation of (+)-valencene by recombinant *E. coli* expressing CYP109B1 from *Bacillus subtilis* in a two-liquid-phase system. Microb Cell Fact 8:36.

Girhard M, Schuster S, Dietrich M, Durre P, Urlacher VB (2007) Cytochrome P450 monooxygenase from *Clostridium acetobutylicum*: a new α-fatty acid hydroxylase. Biochem Biophys Res Commun 362:114-119.

Goni G, Zollner A, Lisurek M, Velazquez-Campoy A, Pinto S, Gomez-Moreno C, Hannemann F, Bernhardt R, Medina M (2009) Cyanobacterial electron carrier proteins as electron donors to CYP106A2 from *Bacillus megaterium* ATCC 13368. Biochim Biophys Acta 1794:1635-1642.

Green AJ, Munro AW, Cheesman MR, Reid GA, von Wachenfeldt C, Chapman SK (2003) Expression, purification and characterisation of a *Bacillus subtilis* ferredoxin: a potential electron transfer donor to cytochrome $P450_{Biol}$. J Inorg Biochem 93:92-99.

Gustafsson MC, Roitel O, Marshall KR, Noble MA, Chapman SK, Pessegueiro A, Fulco AJ, Cheesman MR, von Wachenfeldt C, Munro AW (2004) Expression, purification, and characterization of *Bacillus subtilis* cytochromes P450 CYP102A2 and CYP102A3: flavocytochrome homologues of $P450_{BM3}$ from *Bacillus megaterium*. Biochemistry 43:5474-5487.

Hannemann F, Bera AK, Fischer B, Lisurek M, Teuchner K, Bernhardt R (2002) Unfolding and conformational studies on bovine adrenodoxin probed by engineered intrinsic tryptophan fluorescence. Biochemistry 41:11008-11016.

Hannemann F, Bichet A, Ewen KM, Bernhardt R (2007) Cytochrome P450 systems--biological variations of electron transport chains. Biochim Biophys Acta 1770:330-344.

Hannemann F, Virus C, Bernhardt R (2006) Design of an *Escherichia coli* system for whole cell mediated steroid synthesis and molecular evolution of steroid hydroxylases. J Biotechnol 124:172-181.

Huang JJ, Kimura T (1973) Studies on adrenal steroid hydroxylases. Oxidation-reduction properties of adrenal iron-sulfur protein (adrenodoxin). Biochemistry 12:406-409.

Jenkins CM, Waterman MR (1998) NADPH-flavodoxin reductase and flavodoxin from *Escherichia coli*: characteristics as a soluble microsomal P450 reductase. Biochemistry 37:6106-6113.

Kunst F, Ogasawara N, Moszer I, Albertini AM, Alloni G, Azevedo V, Bertero MG, Bessieres P, Bolotin A, Borchert S, Borriss R, Boursier L, Brans A, Braun M, Brignell SC, Bron S, Brouillet S, Bruschi CV, Caldwell B, Capuano V, Carter NM, Choi SK, Codani JJ, Connerton IF, Danchin A, et al. (1997) The complete genome sequence of the gram-positive bacterium *Bacillus subtilis*. Nature 390:249-256.

Lawson RJ, Leys D, Sutcliffe MJ, Kemp CA, Cheesman MR, Smith SJ, Clarkson J, Smith WE, Haq I, Perkins JB, Munro AW (2004a) Thermodynamic and biophysical characterization of cytochrome P450$_{Biol}$ from *Bacillus subtilis*. Biochemistry 43:12410-12426.

Lawson RJ, von Wachenfeldt C, Haq I, Perkins J, Munro AW (2004b) Expression and characterization of the two flavodoxin proteins of *Bacillus subtilis*, YkuN and YkuP: biophysical properties and interactions with cytochrome P450$_{Biol}$. Biochemistry 43:12390-12409.

Mandai T, Fujiwara S, Imaoka S (2009) A novel electron transport system for thermostable CYP175A1 from *Thermus thermophilus* HB27. FEBS J 276:2416-2429.

Matsuzaki F, Wariishi H (2004) Functional diversity of cytochrome P450s of the white-rot fungus *Phanerochaete chrysosporium*. Biochem Biophys Res Commun 324:387-393.

Maurer S, Urlacher V, Schulze H, Schmid RD (2003) Immobilisation of P450$_{BM3}$ and an NADP$^+$ cofactor recycling system: towards a technical application of heme-containing monooxygenases in fine chemical synthesis. Adv Synth Catal 345:802-810.

McIver L, Leadbeater C, Campopiano DJ, Baxter RL, Daff SN, Chapman SK, Munro AW (1998) Characterisation of flavodoxin NADP$^+$ oxidoreductase and flavodoxin; key components of electron transfer in *Escherichia coli*. Eur J Biochem 257:577-585.

Meyer K (2002) Colorful antioxidants - Carotenoids: Significance and technical syntheses. Chemie unserer Zeit 36:178-192.

Miura Y, Fulco AJ (1975) Omega-1, Omega-2 and Omega-3 hydroxylation of long-chain fatty acids, amides and alcohols by a soluble enzyme system from *Bacillus megaterium*. Biochim Biophys Acta 388:305-317.

Nelson DR (2006) Cytochrome P450 nomenclature, 2004. Methods Mol Biol 320:1-10.

Omura T, Sato R (1964a) The Carbon Monoxide-Binding Pigment of Liver Microsomes. I. Evidence for Its Hemoprotein Nature. J Biol Chem 239:2370-2378.

Omura T, Sato R (1964b) The Carbon Monoxide-Binding Pigment of Liver Microsomes. II. Solubilization, Purification, and Properties. J Biol Chem 239:2379-2385.

Purdy MM, Koo LS, Ortiz de Montellano PR, Klinman JP (2004) Steady-state kinetic investigation of cytochrome P450$_{cam}$: interaction with redox partners and reaction with molecular oxygen. Biochemistry 43:271-281.

Sagara Y, Wada A, Takata Y, Waterman MR, Sekimizu K, Horiuchi T (1993) Direct expression of adrenodoxin reductase in *Escherichia coli* and the functional characterization. Biol Pharm Bull 16:627-630.

Sambrook J, Russell D (2001) Molecular Cloning: A Laboratory Manual. Cold Spring Harbor Laboratory Press, New York.

Schenkman JB, Sligar SG, Cinti DL (1981) Substrate interaction with cytochrome P450. Pharmacol Ther 12:43-71.

Sevrioukova I, Truan G, Peterson JA (1996) The flavoprotein domain of $P450_{BM3}$: expression, purification, and properties of the flavin adenine dinucleotide- and flavin mononucleotide-binding subdomains. Biochemistry 35:7528-7535.

Sirim D, Wagner F, Lisitsa A, Pleiss J (2009) The cytochrome P450 engineering database: Integration of biochemical properties. BMC Biochem 10:27.

Sowden RJ, Yasmin S, Rees NH, Bell SG, Wong LL (2005) Biotransformation of the sesquiterpene (+)-valencene by cytochrome $P450_{cam}$ and $P450_{BM3}$. Org Biomol Chem 3:57-64.

Uhlmann H, Kraft R, Bernhardt R (1994) C-terminal region of adrenodoxin affects its structural integrity and determines differences in its electron transfer function to cytochrome P450. J Biol Chem 269:22557-22564.

Urlacher VB, Makhsumkhanov A, Schmid RD (2006) Biotransformation of β-ionone by engineered cytochrome $P450_{BM3}$. Appl Microbiol Biotechnol 70:53-59.

Virus C, Bernhardt R (2008) Molecular evolution of a steroid hydroxylating cytochrome P450 using a versatile steroid detection system for screening. Lipids 43:1133-1141.

Virus C, Lisurek M, Simgen B, Hannemann F, Bernhardt R (2006) Function and engineering of the 15-β-hydroxylase CYP106A2. Biochem Soc Trans 34:1215-1218.

Wang ZQ, Lawson RJ, Buddha MR, Wei CC, Crane BR, Munro AW, Stuehr DJ (2007) Bacterial flavodoxins support nitric oxide production by *Bacillus subtilis* nitric-oxide synthase. J Biol Chem 282:2196-2202.

2.1.1.9. Supplementary Material

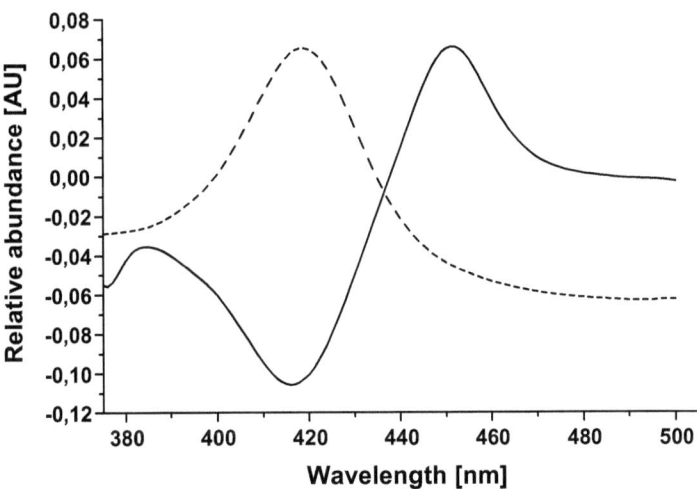

Figure 2-6: Spectra of CYP109B1 in its different redox states. Redox states are labeled as follows: oxidized (dashed line), CO-difference spectrum (solid line).

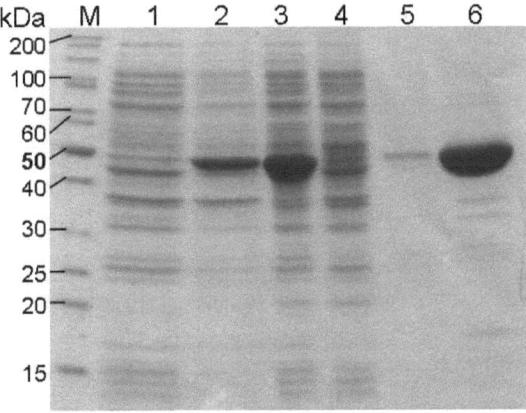

Figure 2-7: 12% SDS-PAGE of CYP109B1 expression in recombinant *E. coli* and purification by immobilized metal affinity chromatography (IMAC). Non-induced *E. coli* whole-cells (line 1); IPTG-induced *E. coli* whole-cells (line 2); soluble protein fraction after cell lysis (line 3); flow-through protein fraction of IMAC purification (line 4); washed out protein fraction of IMAC purification (line 5); purified CYP109B1 (line 6); PageRuler[TM] Unstained Protein Ladder (line M). The molecular weight of CYP109B1 is around 45.0 kDa.

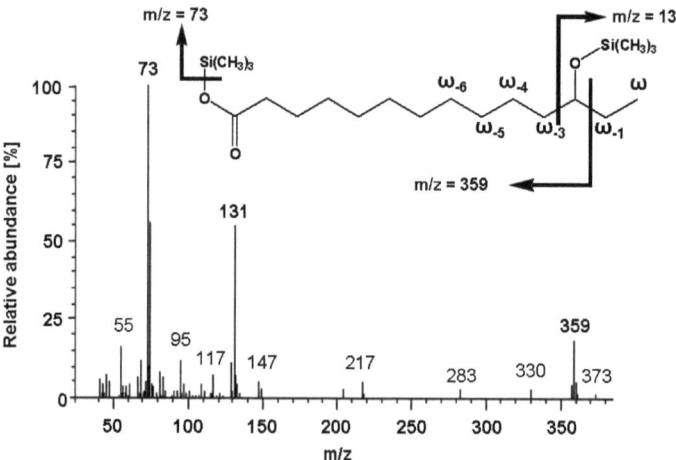

Figure 2-8: Example for the identification of fatty acid oxidation products through their mass fragmentation pattern after derivatization with trimethylchlorosilane. The characteristic mass fragments for the ω_{-2}-hydroxylated product of myristic acid **5** are 73, 131 and 359, respectively.

Table 2-5: Product distribution and conversion of saturated and esterified fatty acids by CYP109B1.

Type of substrate	Chain length (**No.**)	Conversion [%]	Regioselectivity [%][a]							
			ω$_{-8}$	ω$_{-7}$	ω$_{-6}$	ω$_{-5}$	ω$_{-4}$	ω$_{-3}$	ω$_{-2}$	ω$_{-1}$
Saturated fatty acids	8 (**1**)	0.0	—	—	—	—	—	—	—	—
	10 (**2**)	3.3 ± 0.2	—	—	—	—	—	—	—	100
	12 (**3**)	95.0 ± 2.6	—	—	—	0.4	2.7	16.1	37.5	43.5
	13 (**4**)	99.1 ± 0.1	—	—	0.5	1.1	9.3	18.5	39.8	30.8
	14 (**5**)	99.1 ± 0.3	—	—	0.7	11.6	19.1	25.8	26.1	16.7
	15 (**6**)	98.2 ± 0.4	—	0.3	5.0	17.9	36.4	20.8	14.7	5.1
	16 (**7**)	50.6 ± 4.7	2.6	7.3	12.3	23.4	29.5	12.7	7.9	4.5
	18 (**8**)	2.9 ± 0.3	(+)	(+)	(+)	(+)	(+)	(+)	—	—
Saturated fatty acid methylesters	12 (**9**)	77.6 ± 7.6	—	—	—	0.6	2.8	14.5	38.9	43.3
	14 (**10**)	79.8 ± 8.3	—	—	1.1	9.6	21.3	27.9	25.0	15.3
	16 (**11**)	30.1 ± 2.9	1.1	7.8	8.2	18.8	21.8	16.9	15.7	9.8
Saturated fatty acid ethylesters	12 (**12**)	31.6 ± 0.7	—	—	—	2.6	4.3	12.9	39.8	40.4
	14 (**13**)	33.7 ± 3.1	—	—	4.0	4.7	20.9	30.9	23.7	16.0

[a] Values are given in % of the total product as mean of n ≥ 3
— Compound was not observed or below the detection limits (0.6 µmol l^{-1})
(+) Compound was observed, but quantitative analysis could not be done due to low conversion of the substrate

2.1.2. Manuscript: Regioselective biooxidation of (+)-valencene by recombinant *E. coli* expressing CYP109B1 from *Bacillus subtilis* in a two liquid phase system

Material from this chapter appears in:
Marco Girhard[a,b], Kazuhiro Machida[b], Masashi Itoh[b], Rolf D. Schmid[a], Akira Arisawa[b] and Vlada B. Urlacher[a,*], 2009, Regioselective biooxidation of (+)-valencene by recombinant *E. coli* expressing CYP109B1 from *Bacillus subtilis* in a two liquid phase system, *Microbial Cell Factories* 8:36.

[a] Institute of Technical Biochemistry, Universität Stuttgart, 70569 Stuttgart, Germany
[b] Bioresource Laboratories, Mercian Corporation, 1808 Nakaizumi, Iwata, Shituoka 438-0078, Japan
* Corresponding author

This is an Open Access article distributed under the terms of the Creative Commons Attribution License (http://creativecommons.org/licenses/by/2.0).
The original manuscript is available online at: http://www.microbialcellfactories.com

Microbial Cell Factories

Research

Open Access

Regioselective biooxidation of (+)-valencene by recombinant *E. coli* expressing CYP109B1 from *Bacillus subtilis* in a two-liquid-phase system

Marco Girhard[1,2], Kazuhiro Machida[2], Masashi Itoh[2], Rolf D Schmid[1], Akira Arisawa[2] and Vlada B Urlacher*[1]

Address: [1]Institute of Technical Biochemistry, Universitaet Stuttgart, Allmandring 31, 70569 Stuttgart, Germany and [2]Bioresource Laboratories, Mercian Corporation, 1808 Nakaizumi, Iwata, Shizuoka 438-0078, Japan

Email: Marco Girhard - marco.girhard@itb.uni-stuttgart.de; Kazuhiro Machida - machida-k@mercian.co.jp; Masashi Itoh - ito-ms@mercian.co.jp; Rolf D Schmid - rolf.d.schmid@itb.uni-stuttgart.de; Akira Arisawa - arisawa-a@mercian.co.jp; Vlada B Urlacher* - itbvur@itb.uni-stuttgart.de

* Corresponding author

Published: 10 July 2009

Microbial Cell Factories 2009, **8**:36 doi:10.1186/1475-2859-8-36

Received: 12 February 2009
Accepted: 10 July 2009

This article is available from: http://www.microbialcellfactories.com/content/8/1/36

© 2009 Girhard et al; licensee BioMed Central Ltd.
This is an Open Access article distributed under the terms of the Creative Commons Attribution License (http://creativecommons.org/licenses/by/2.0), which permits unrestricted use, distribution, and reproduction in any medium, provided the original work is properly cited.

Abstract

Background: (+)-Nootkatone (**4**) is a high added-value compound found in grapefruit juice. Allylic oxidation of the sesquiterpene (+)-valencene (**1**) provides an attractive route to this sought-after flavoring. So far, chemical methods to produce (+)-nootkatone (**4**) involve unsafe toxic compounds, whereas several biotechnological approaches applied yield large amounts of undesirable byproducts. In the present work 125 cytochrome P450 enzymes from bacteria were tested for regioselective oxidation of (+)-valencene (**1**) at allylic C2-position to produce (+)-nootkatone (**4**) via *cis*- (**2**) or *trans*-nootkatol (**3**). The P450 activity was supported by the coexpression of putidaredoxin reductase (PdR) and putidaredoxin (Pdx) from *Pseudomonas putida* in *Escherichia coli*.

Results: Addressing the whole-cell system, the cytochrome CYP109B1 from *Bacillus subtilis* was found to catalyze the oxidation of (+)-valencene (**1**) yielding nootkatol (**2** and **3**) and (+)-nootkatone (**4**). However, when the *in vivo* biooxidation of (+)-valencene (**1**) with CYP109B1 was carried out in an aqueous milieu, a number of undesired multi-oxygenated products has also been observed accounting for approximately 35% of the total product. The formation of these byproducts was significantly reduced when aqueous-organic two-liquid-phase systems with four water immiscible organic solvents – isooctane, *n*-octane, dodecane or hexadecane – were set up, resulting in accumulation of nootkatol (**2** and **3**) and (+)-nootkatone (**4**) of up to 97% of the total product. The best productivity of 120 mg l⁻¹ of desired products was achieved within 8 h in the system comprising 10% dodecane.

Conclusion: This study demonstrates that the identification of new P450s capable of producing valuable compounds can basically be achieved by screening of recombinant P450 libraries. The biphasic reaction system described in this work presents an attractive way for the production of (+)-nootkatone (**4**), as it is safe and can easily be controlled and scaled up.

Background

The sesquiterpenes are a large class of terpenoid compounds and common constituents of plant essential oils. The parent sesquiterpenes are often readily available and their oxygenation gives derivatives which are sought-after fragrances and flavorings, pharmaceuticals or building blocks for chemical synthesis [1]. For example, (+)-valencene (1) is found in citrus oils and can be cheaply obtained from oranges. Recently the valencene synthase gene from orange has been cloned and functionally expressed in E. coli [2,3]. The selective oxidation of (+)-valencene (1) at allylic C2-position yields cis- (2) and trans-nootkatol (3), which can be further oxidized to (+)-nootkatone (4), a high added-value commercial flavoring (Figure 1). Traditionally nootkatone is extracted from grapefruits, and its price and availability is dependent on the annual harvest, which is restricted to a narrow producing area and very sensitive to weather conditions. Thus,

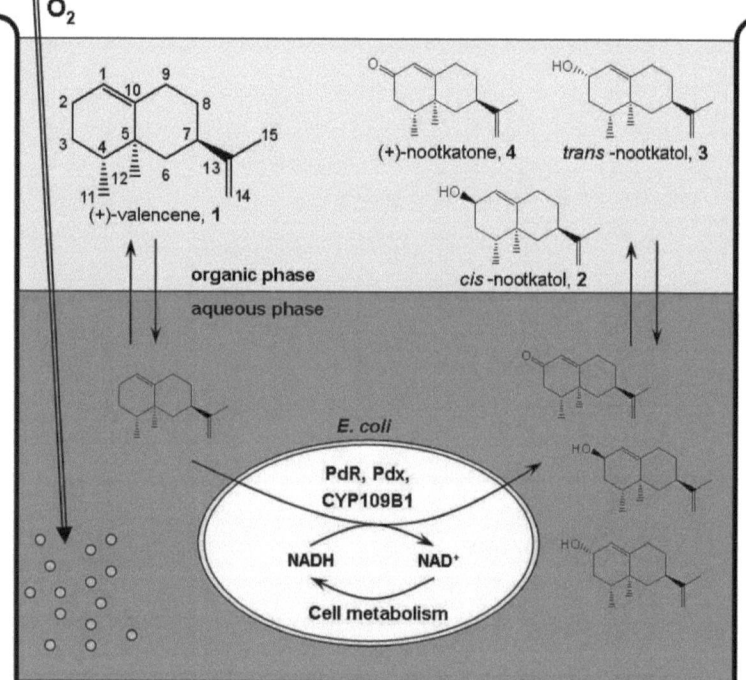

Figure 1
Schematic experimental setup for the conversion of (+)-valencene (1) by CYP109B1 in a biphasic system. (+)-Valencene (1) is converted to cis-nootkatol (2), trans-nootkatol (3) and (+)-nootkatone (4) by recombinant E. coli expressing CYP109B1, putidaredoxin reductase (PdR) and putidaredoxin (Pdx). The organic phase acts as substrate reservoir and allows the accumulation of the mono-oxygenated conversion products, which prevents them from overoxidation.

various approaches for chemical or biotechnological production of (+)-nootkatone have been investigated in the past years (very recently reviewed by Fraatz et al. [4]).

Chemical synthesis of (+)-nootkatone (4) from (+)-valencene (1) has been studied with *tert*-butyl chromate [5], via copper(I)-mediated oxidation by alkyl hydroperoxides [6] and with surface-functionalized silica supported by metal catalysts such as Co^{2+} and Mn^{2+} [7]. However, these synthetic methods are neither safe nor environment-friendly, because they involve toxic heavy metals or peroxides.

Biotechnological processes to achieve this oxidation have also been designed. Recent attempts include the manufacturing of (+)-nootkatone (4) with green algae like *Chlorella* or *Euglena* [8], fungi such as *Aspergillus niger, Fusarium culmorum* [9], *Mucor* sp. [10,11] or the ascomycete *Chaetomium globosum* [12]. Cell free enzymatic reactions for the conversion of (+)-valencene (1) exploiting enzymes from *Cichorium intybus* L. roots [13], lignin peroxidase [14] and fungal laccase [15] have also been reported.

Several cytochrome P450 monooxygenases (P450s or CYPs) have been reported to catalyze terpene oxidation [16-21]. Cytochrome P450s belong to a superfamily of heme b-containing proteins that accept a vast number of organic compounds and catalyze an enormous variety of oxidative reactions. These reactions include hydroxylation, epoxidation, heteroatom oxidation, C-C bond cleavage via successive hydroxylations and many other complex reactions [22,23].

Interestingly, (+)-nootkatone (4) has been shown to inhibit activity of some human P450s, including CYP2A6 and CYP2C19 [24]. However, given the large superfamily of these enzymes (> 8100 sequences, http://drnelson.utmem.edu/p450stats.Feb2008.htm) and their diverse substrate range, it is unlikely that (+)-nootkatone (4) will inhibit all P450s. Indeed, a recent report describes the hydroxylation of (+)-valencene (1) by mutants of the cytochromes $P450_{cam}$ from *Pseudomonas putida* and $P450_{BM3}$ from *Bacillus megaterium* [25]. In both cases, however, certain improvements are required. While the reported $P450_{cam}$-mutants demonstrated quite low activity, the $P450_{BM3}$-mutants, although being active, had low chemo- and regioselectivity and produced up to six byproducts besides nootkatol (2 and 3) and (+)-nootkatone (4). Another report described a membrane-bound plant P450 monooxygenase from *Hyoscyamus muticus*, which was able to oxidize (+)-valencene (1) to nootkatol [16]. Generally, all current biotechnological systems for production of (+)-nootkatone (4) have in common an unsatisfactory accumulation of the C2-oxidized compounds because of low enzyme activity and/or the drawback of the formation of large amounts of byproducts.

In the search for new P450 monooxygenases with activity and high regioselectivity towards (+)-valencene (1), we screened 125 P450 enzymes from a novel recombinant P450 library, based on about 250 bacterial cytochrome P450 monooxygenases, co-expressed with putidaredoxin reductase (PdR) and putidaredoxin (Pdx) in *Escherichia coli*. The library contains, for example, the complete P450 complement of *Streptomyces coelicolor* A3(2), *Bacillus subtilis* 168, *Nocardia farcinica* IFM 10152 and *Bradyrhizobium japonicum* USDA110. Generally, about 70% of the P450 candidates originate from actinomycetes. Construction and application of this P450 library has recently been reported with respect to steroid oxidation [26].

The actual screening identified CYP109B1 from *Bacillus subtilis*, which catalyzed the regioselective oxidation of (+)-valencene (1) to nootkatol (2 and 3) and (+)-nootkatone (4). When the whole-cell biooxidation of (+)-valencene (1) with *E. coli* expressing CYP109B1 and electron transfer partners was carried out in an aqueous milieu, a number of undesired multi-oxygenated oxidation products was observed. In order to reduce byproduct-formation biphasic systems with water immiscible organic solvents were applied and the best system was scaled-up.

Results and discussion
Screening of the recombinant P450 library
In our approach to find new P450s capable to oxidize (+)-valencene (1) regioselectively at allylic C2-position, 125 bacterial P450s from an *E. coli*-based recombinant P450 library were screened. Generally, P450 monooxygenases require electron transfer partners for activity. Since natural electron transfer partners for most of these 125 P450s are unknown, PdR and Pdx from *P. putida* were co-expressed in the *E. coli* cells. In the screening two enzymes were found to be able to oxidize (+)-valencene (1), namely CYP109B1 from *B. subtilis* 168 and CYP105 from *Nonomuraea recticatena* NBRC 14525 (also referred to as P450MoxA [27]). The conversion of (+)-valencene (1) by CYP109B1 from *B. subtilis* resulted in six oxidation products, with the desired mono-oxygenated compounds nootkatol (2 and 3) and (+)-nootkatone (4) accounting for approximately 65% of the total product, whereas oxidation of (+)-valencene (1) by P450MoxA yielded 15 different mono- and multi-oxygenated oxidation products (which mostly could not be identified), with the desired products accounting for less than 10% of the total product (Figure 2). The *E. coli* BL21(DE3) cells expressing only

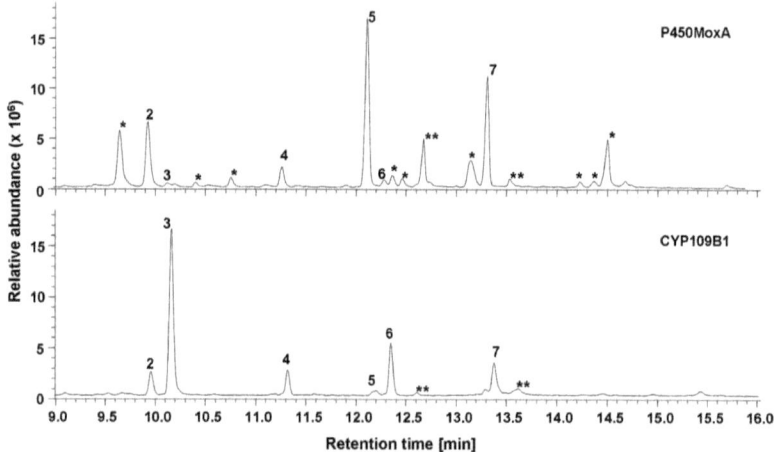

Figure 2
GC diagrams of extracts of (+)-valencene (1) conversions. Conversion of 2 mM (+)-valencene (1) was carried out with recombinant E. coli expressing PdR, Pdx and either P450MoxA or CYP109B1 in 2 ml aqueous CV2 buffer with 2% DMSO for 8 h at 30°C and extracted with ethyl acetate. Numbered peaks represent cis-nootkatol (2), trans-nootkatol (3), (+)-nootkatone (4) and overoxidation products (5, 6 and 7). * Conversion products of (+)-valencene (1) that could not be identified. ** Non-specific peaks that were also present in the negative control.

PdR and Pdx did not convert (+)-valencene. Because of its high regioselectivity, CYP109B1 was chosen as biocatalyst for further studies.

Biooxidation of (+)-valencene by CYP109B1 in an aqueous system

Whole-cell biotransformation with E. coli BL21(DE3) cells expressing PdR-, Pdx- and CYP109B1 was first carried out in aqueous milieu (CV2 buffer, see Material and Methods). The average expression level of CYP109B1 verified through CO-difference spectra was 160 mg l⁻¹, which corresponds to 7.6 mg g⁻¹ cell wet weight (cww). Since in all experiments 70 g cww l⁻¹ were used, the final P450 concentration was approximately 530 μg ml⁻¹. Expression of PdR and Pdx was verified by sodium dodecyl sulphate polyacrylamid gel electrophoresis (SDS-PAGE) [see Additional file 1].

When (+)-valencene (1) was added directly to the cell suspension, the oxidation of the substrate by CYP109B1 proceeded with 270 nM min⁻¹ and after 8 h 6% conversion was achieved. Because of the low conversion, 2% dimethyl sulfoxid (DMSO) was added to the reaction. DMSO increases the solubility of (+)-valencene (1), which is poorly soluble in the aqueous reaction medium. In consequence the volumetric productivity could be increased up to 990 nM min⁻¹ and after 8 h 25% of the substrate was converted.

Electron transfer to CYP109B1 in E. coli cells can occur either via co-expressed heterologuos PdR and Pdx and/or via endogenous flavo- or ferredoxins and reductases from the host strain. To investigate, if co-expressed PdR and Pdx in fact support the activity of CYP109B1, E. coli BL21(DE3) was transformed with the pET28a-CYP109B1-plasmid comprising exclusively the CYP109B1-gene without PdR and Pdx (see Material and Methods section). In this case the biooxidation of (+)-valencene (1) preceded with only 40 nM min⁻¹, regardless of the addition of DMSO, which demonstrates that co-expression of PdR and Pdx is necessary for activity of CYP109B1 in vivo.

For further evaluation of the electron transport to CYP109B1 via PdR and Pdx, a reconstituted in vitro system with recombinantly expressed and purified CYP109B1, PdR and Pdx [see Additional file 1] was set up as described in the material and methods section. In this system, the NADH oxidation rate was 64 nmol nmol^{-1} P450 min^{-1} and approximately 10% conversion of 200 µM (+)-valencene (1) was observed, however, only if all three proteins were added to the reaction mixture. Mixtures without PdR and/or Pdx did not show any activity towards (+)-valencene (1) (data not shown). Detailed biochemical characterization of CYP109B1 is a topic of our current investigations.

Gas chromatography (GC) analysis of reaction mixtures after the *in vivo* (+)-valencene (1) biotransformation revealed several new peaks besides the substrate peak (Figure 2). According to their mass fragmentation spectra (MS) and by comparison with authentic reference compounds three peaks were identified as *cis*-nootkatol (2) (retention time [RT] = 9.9 min), *trans*-nootkatol (3) (RT = 10.1 min) and nootkatone (4) (RT = 11.3 min) accounting for 6.2, 55.1 and 4.0% of the total product, respectively. Three other products were observed at RTs of 12.1, 12.3 and 13.3 min accounting for 2.3, 27.2 and 5.2% of the total product. A time course of the reaction demonstrated that these byproducts appeared shortly after the two nootkatol-isomers (2 and 3) and nootkatone (4) were formed, which indicates that these compounds could represent products of overoxidation (data not shown). Furthermore, the fragmentation spectra and molecular weights of these compounds indicated the incorporation of two oxygen atoms into (+)-valencene (1) [see Additional file 2]. In a control experiment with nootkatone (4) as substrate instead of (+)-valencene (1), only the product with a RT of 13.3 min was identified by GC/MS analysis (data not shown). Obviously, this product is the oxidized nootkatone. Accordingly, when a mixture of nootkatol isomers (2 and 3) was added as substrate, the products with RT of 12.1 and 12.3 min and small amounts of (+)-nootkatone (4) were observed. Thus these products were assumed to be products of further oxidation of the two nootkatol isomers (2 and 3), which is in agreement with their MS data. In order to investigate if the primary oxidation products could be protected from overoxidation, we applied an aqueous-organic two-liquid-phase approach as shown in figure 1.

Remarkably, in the control experiment with nootkatone (4) instead of (+)-valencene (1), CYP109B1 oxidized approximately 60% of (+)-nootkatone (4) (final concentration 2 mM) within 2 h of conversion, whereas (+)-valencene (1) (final concentration 2 mM) was oxidized by only approximately 5% within this time. This means that nootkatone is actually a better substrate for CYP109B1 than valencene. Interestingly, this phenomena has also been reported for the P450$_{BM3}$ wild type enzyme and mutants, which accept nootkatone as substrate much better than valencene [25].

Biocompatibility of organic solvents

Aqueous-organic two-liquid-phase systems (further referred to as biphasic systems) represent a powerful biotechnological tool for biotransformations of toxic organic compounds [28-30]. This approach allows 1) the use of high overall concentrations of hydrophobic toxic substrates by regulating substrate and product concentrations in the aqueous biocatalyst phase, and 2) simple *in situ* product recovery into organic phase. Four water-immiscible organic solvents (*n*-octane, isooctane, dodecane and hexadecane) were chosen to set up biphasic systems and compared to the aqueous system. Control experiments showed that none of the organic solvents was oxidized by the host strain.

It is known that some organic solvents used as a second phase can damage the microbial cells [31]. Viability assays conducted after 8 h of (+)-valencene (1) conversion with *E. coli* expressing PdR, Pdx and CYP109B1 in systems with 20% (v/v) solvent showed that isooctane and octane had a strong effect on cell viability, with only 20% of cells surviving, whereas the more hydrophobic solvents dodecane and hexadecane were totally biocompatible with *E. coli*, as they did not harm the cell viability considerably. The addition of 2% DMSO had no considerable effect on cell viability in aqueous systems and in systems with dodecane or hexadecane, but led to further diminution of cell viability with octane and isooctane (Figure 3).

Biooxidation of (+)-valencene (1) in biphasic systems

First, biphasic systems without addition of DMSO were set up. In these systems, however, the volumetric productivities were low (≤ 100 nM min^{-1}) and after 8 h only 2% or less (+)-valencene (1) conversion was achieved in all systems tested. Presumably, the substrate holding capacity of the organic solvents was too high, preventing its oxidation in the aqueous phase. High distribution coefficients (log D) measured for (+)-valencene (1) in all four organic solvents confirm this observation (Table 1).

After addition of 2% DMSO the conversion of (+)-valencene (1) rose significantly in all biphasic systems employed, but was still lower in comparison to the aqueous system where 25% conversion was reached (Table 2). Generally, the addition of an organic phase led to a strong reduction of the generated multi-oxygenated byproducts causing a shift of the product distribution towards the mono-oxygenated compounds nootkatol (2 and 3) and (+)-nootkatone (4) (Table 2). Moreover, the amount of byproducts decreased with increasing amounts of organic solvents from 10 to 20%. This tendency was similar for all four solvents tested. Basically, addition of 10% organic

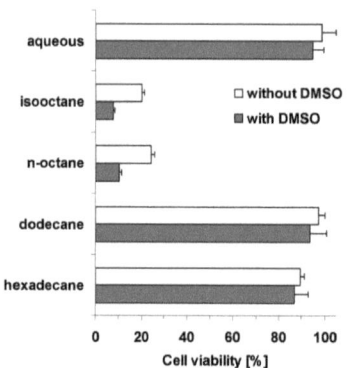

Figure 3
Cell viability of *E. coli*. Cell viability of recombinant *E. coli* expressing PdR, Pdx and CYP109B1 was measured after 8 h exposure to biphasic systems consisting of CV2 buffer with or without 2% DMSO and 20% (v/v) organic solvents and compared to pure CV2 buffer (aqueous). Values represent mean ± standard deviation.

solvent reduced the amount of byproducts to less than 12% of the total product; 20% organic phase resulted in less than 7% byproducts. The best result in terms of reduction of byproducts was achieved with 20% hexadecane, were nootkatol (2 and 3) and (+)-nootkatone (4) accounted for 97% of the total product (Figure 4).

Generally, the volumetric productivities of the biphasic systems were lower than those of the aqueous system (990 nM min^{-1}). For example, addition of 10% of isooctane resulted in a 60% loss in original volumetric productivity (400 nM min^{-1}), 20% of isooctane – in an almost 70% loss (Table 2). This high reduction is obviously due to the toxicity of isooctane to *E. coli*, as described in the previous section. For *n*-octane – which is toxic for *E. coli* as well – volumetric productivities of 770 and 502 nM min^{-1} were obtained at 10% or 20% *n*-octane, respectively. Dodecane had the slightest effect on volumetric productivity; about 87% could be retained in the system with 10% dodecane after 8 h conversion (860 nM min^{-1}). Interestingly, the retained volumetric productivities with hexadecane were significantly lower then those observed in the presence of *n*-octane, especially at 20% (Table 2), although the viabil-

ity assay demonstrated that hexadecane is benign for *E. coli*.

When we determined the log D of (+)-valencene (1), *trans*-nootkatol (3) and (+)-nootkatone (4) for the biphasic systems with 20% organic solvent in the presence of 2% DMSO (Table 1), in the biphasic system with hexadecane a log D of 0.62 for (+)-valencene (1) was obtained, which is two-fold higher than with dodecane (0.32) and 3.5-times higher than with *n*-octane (0.18). The high log D for hexadecane implies a larger substrate holding capacity of this solvent, which might explain the significantly reduced volumetric productivity of the whole-cell process for this solvent compared to dodecane. The highest log D for (+)-valencene under these conditions was determined in isooctane (0.87), presumably also contributing to the low volumetric productivities in addition to the toxicity of this solvent.

Taking into account both, the reduction of byproduct formation and volumetric productivities, the best performance was observed in the biphasic system comprising 10% dodecane and 2% DMSO.

Reaction scale-up

Conversion of (+)-valencene (1) was carried out in 20 ml volume for the biphasic system comprising 10% dodecane and 2% DMSO. The influence of cell density and substrate concentration on the yield of nootkatol (2 and 3) and (+)-nootkatone (4) was investigated.

At a constant cww of 70 g l^{-1} (corresponding to 18.4 g l^{-1} cell dry weight (cdw)) higher substrate concentrations resulted, as expected, in higher overall amounts of desired products, however the substrate conversion values decreased (Table 3). For example, from 409 mg l^{-1} (+)-valencene (1) 61.6 mg l^{-1} nootkatol (2 and 3) and (+)-nootkatone (4) were produced, and from 818 mg l^{-1} (+)-valencene (1) – 96.0 mg l^{-1} (Table 3). Product distribution did not change, independently on substrate concentration.

At a constant substrate concentration of 409 mg l^{-1} and two-fold increased cell density (140 g l^{-1}) the yield of desired products could be increased from 61.6 mg to 89.1 mg l^{-1}. A four-fold higher cell density resulted in 94.2 mg l^{-1} (Table 3). Thus, higher cell densities at a constant substrate concentration lead to higher substrate conversion and higher overall amounts of nootkatol (2 and 3) and (+)-nootkatone (4), but lower product yields per g of cells.

Under optimized reaction conditions 120 mg l^{-1} of desired products were produced. In comparison, the con-

Table 1: Distribution coefficients for (+)-valencene (1), trans-nootkatol (3) and (+)-nootkatone (4)

Compound	DMSO [%][a]	Distribution coefficient (log D)[b]			
		isooctane	n-octane	dodecane	hexadecane
(+)-valencene (1)	0	2.79 ± 0.03	2.61 ± 0.04	2.74 ± 0.05	2.62 ± 0.10
trans-nootkatol (3)	0	2.43 ± 0.03	2.31 ± 0.04	2.12 ± 0.02	2.11 ± 0.04
(+)-nootkatone (4)	0	2.63 ± 0.04	2.55 ± 0.03	2.68 ± 0.08	2.37 ± 0.12
(+)-valencene (1)	2	0.87 ± 0.05	0.18 ± 0,03	0.32 ± 0.08	0.62 ± 0.02
trans-nootkatol (3)	2	1.55 ± 0.05	1.57 ± 0.05	1.39 ± 0.06	1.23 ± 0.04
(+)-nootkatone (4)	2	1.77 ± 0.04	1.98 ± 0.05	1.83 ± 0.07	1.29 ± 0.06

[a]Final concentration in the mixture.
[b]Values represent mean ± standard deviation.

centration of desired products in the aqueous system under the same conditions reached 83 mg l⁻¹.

Conclusion

It was demonstrated that the identification of new P450s capable of producing valuable compounds can basically be achieved by the screening of recombinant P450 libraries. The P450-library used in this study is easy to handle and allows high throughput screening of various substrates.

For the biooxidation of (+)-valencene (1) with *E. coli* expressing recombinant CYP109B1 from *Bacillus subtilis*, Pdx and PdR the use of organic solvents improved the accumulation of nootkatol (2 and 3) and (+)-nootkatone (4). Although the addition of organic solvents reduced the volumetric productivity of the whole-cell process, the production of nootkatol is promising. Considering the accumulation of nootkatol (2 and 3) it should be noted that a bottleneck reaction seems to be the oxidation of nootkatol (2 and 3) to (+)-nootkatone (4). Here both, process and enzyme engineering could help to overcome this limitation. For example, CYP109B1 could be optimized for higher affinity and activity towards (+)-valencene (1) and nootkatol (2 and 3) with the goal to achieve an increased productivities of the whole-cell process and higher amounts of (+)-nootkatone (4). Alternatively, a specific dehydrogenase could be expressed in the same *E. coli* host to oxidize nootkatol (2 and 3) to (+)-nootkatone (4) and thereby regenerate NADH, which in addition would reduce metabolic stress for the host strain.

Methods

Enzymes and materials

Restriction endonucleases, T4 DNA ligase, *Pfu* DNA polymerase and isopropyl β-D-thiogalactopyranoside (IPTG) were obtained from Fermentas (St. Leon-Rot, Germany). (+)-Valencene and (+)-nootkatone were from

Table 2: Product distribution and conversion of (+)-valencene (1) by CYP109B1 in biphasic systems[a]

Biphasic system	cis-nootkatol (2)[b]	trans-nootkatol (3)[b]	(+)-nootkatone (4)[b]	RT 12.1 min[b]	RT 12.3 min[b]	RT 13.3 min[b]	Conversion [%][c]	Volumetric productivity [nM min⁻¹]
aqueous	6.2	55.1	4.0	2.3	27.2	5.2	24.7 ± 3.8	990 ± 158
10% dodecane	7.4	82.8	2.6	---	6.1	1.1	19.5 ± 2.9	859 ± 121
20% dodecane	7.6	85.6	2.7	---	3.9	0.2	17.2 ± 1.8	719 ± 75
10% n-octane	7.7	80.2	3.2	0.4	7.3	1.2	16.4 ± 2.7	770 ± 116
20% n-octane	8.5	81.3	3.8	---	5.3	1.1	14.4 ± 2.4	502 ± 101
10% isooctane	10.1	74.9	4.1	0.1	8.5	2.3	9.6 ± 1.6	399 ± 68
20% isooctane	11.3	78.3	4.5	---	4.6	1.3	7.8 ± 1.8	313 ± 84
10% hexadecane	10.8	80.4	2.8	---	5.3	0.7	15.1 ± 3.5	631 ± 150
20% hexadecane	12.6	81.7	3.0	---	2.7	---	9.3 ± 1.2	387 ± 53

[a]Conversion of 2 mM (+)-valencene (1) was carried out with recombinant *E. coli* expressing PdR, Pdx and CYP109B1 in 2 ml CV2 buffer with 2% DMSO for 8 h at 30°C and extracted with ethyl acetate. All values are shown as mean (± standard deviation), n ≥ 3.
[b]Values are given in % of the total product.
[c]Percental amount of (+)-valencene (1) that was converted to products.
RT: Retention time during GC (for unidentified products).
---: Compound was not observed in the reaction. The detection limit was 4 μM.

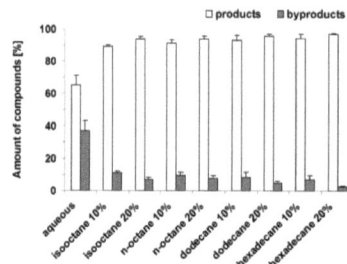

Figure 4
Product distributions of (+)-valencene (1) conversions in biphasic systems. Conversion of 2 mM (+)-valencene (1) was carried out with recombinant *E. coli* expressing PdR, Pdx and CYP109B1 in biphasic systems consisting of 2 ml CV2 buffer, 2% DMSO and organic solvents (v/v) as indicated for 8 h at 30°C and extracted with ethyl acetate. Values represent mean ± standard deviation, n ≥ 3. "Products" sum up the percental amounts of the total product of the two nootkatol isomers (2 and 3) and (+)-nootkatone (4), "byproducts" combine the percental amounts of the total product of the compounds with RT 12.1, 12.3 and 13.3 min, respectively.

Fluka (Buchs, Switzerland). Glucose-6-phosphate dehydrogenase from *Leuconostoc mesenteroides* (EC 1.1.1.49) and other chemicals, solvents and buffer components were purchased from Sigma-Aldrich (Schnelldorf, Germany).

Molecular biological techniques
General molecular biology manipulations and microbiological experiments were carried out by standard methods [32]. Construction of pT7-camAB - a plasmid based on pET11a (Novagen, Darmstadt, Germany) and harboring the *camA* (PdR) and *camB* (Pdx) genes from *Pseudomonas putida* ATCC 17453 - has been described previously [33]. pT7NS-camAB was prepared by insertion of an NdeI-SpeI linker into pT7-camAB and plasmids for the expression of bacterial P450s were created by amplification of annotated P450-encoding genes from genomic DNA of various bacteria as described in [33]. The plasmid for co-expression of PdR, Pdx and CYP109B1 - pCYP109B1-camAB - was constructed by amplification of the CYP109B1-encoding *yjiB*-gene [GenBank:CAB13078] from genomic DNA of the *Bacillus subtilis* strain 168 [34] and by following ligation of the PCR fragment into pT7NS-camAB utilizing the restriction sites for NdeI and SpeI of the NdeI-SpeI linker. pCYP109B1-camAB comprises the genes under control of the IPTG-inducible T7 phage-promoter in an artificial operon in the following order: *yjib* (CYP109B1), *camA* (PdR), *camB* (Pdx).

pET28a-CYP109B1 - a vector for the expression of CYP109B1 alone, without PdR and Pdx - was constructed as follows: The *yjiB*-gene was amplified from pCYP109B1-camAB with the primers 5'-AATA*gctagc*ATGAATGTGT-TAAACCGCCG-3' and 5'-TCG*ctcgag*TTACATTT-TCACACGGAAGC-3'. The PCR was performed with *Pfu* DNA polymerase under the following conditions: 95°C for 4 min, 25 cycles of (95°C for 1 min, 56°C for 30 s, 72°C for 3 min), 72°C for 5 min. The PCR-product was purified, digested with endonucleases NheI and XhoI and inserted into the previously linearized expression vector pET28a(+) (Novagen, Darmstadt, Germany). The resulting DNA-construct encoded for N-terminally His6-tagged CYP109B1 under control of the IPTG-inducible T7 phage-promoter.

pET28a-camA - a vector for the expression of PdR - was constructed by PCR-amplification of the *camA*-gene from pT7NS-camAB with the primers 5'-GGAATTCCATAT-GAACGCAAACGAC-3' and 5'-GATCGAATTCTCAG-GCACTACTCAG-3'. The PCR was done as described for pET28a-CYP109B1, the PCR-product was purified and digested with endonucleases NdeI and EcoRI and inserted into previously linearized expression vector pET28a(+). The resulting plasmid encoded for N-terminally His6-tagged PdR under control of the IPTG-inducible T7 phage-promoter.

The basic steps for the design of pET28a-camB - which incorporates the *camB*-gene under control of the IPTG-inducible T7 phage-promoter for expression of N-terminal His-tagged Pdx - were identical to those for construction of pET28a-camA, except for primers. The primers 5'-GGGAATTCCATATGTCTAAAGTAGTGTAT-3' and 5'-CCGGAATTCTTACCATTGCCTATCGGG-3' were used for PCR in this case.

Protein expression and purification
Recombinant CYP109B1 was expressed under the following conditions: *E. coli* BL21(DE3) cells were transformed with pET28a-CYP109B1 and transformants were selected on LB-agar plates with kanamycin (30 μg ml^{-1}). 400 ml LB supplemented with 30 μg ml^{-1} of kanamycin were than inoculated with 2 ml overnight culture - grown from a single colony - and grown at 37°C and 180 rpm, until the optical density at 600 nm (OD$_{600}$) reached approximately 1.0. 100 μM IPTG, 80 μg ml^{-1} 5-aminolevulinic acid and 0.1 μM FeSO$_4$ were added and the culture was grown for another 19 h at 30°C and 140 rpm. Cells were harvested by centrifugation at 10,000 × g for 20 min, the supernatant was discarded and the cell pellet was resuspended in

Table 3: Yields of nootkatol (2 and 3) and (+)-nootkatone (4) in biphasic systems with 10% dodecane[a]

Cww [g l⁻¹]	c_valencene [mg l⁻¹]	Conversion [%][b]	Yield [mg g⁻¹ cdw]	Yield [mg l⁻¹]
70	409	15.1	3.35	61.6
70	613	13.2	4.40	80.9
70	818	11.8	5.22	96.0
140	409	21.8	2.42	89.1
280	409	23.0	1.28	94.2
140	818	14.1	3.14	118.5

[a]Conversion of (+)-valencene (1) was carried out with recombinant E. coli expressing PdR, Pdx and CYP109B1 in 20 ml CV2 buffer with 2% DMSO for 8 h at 30°C and extracted with ethyl acetate. Cell wet weight (cww) was as indicated; 70 g cww correspond to 18.4 g cell dry weight (cdw). Values are shown as mean of two experiments.
[b]Percental amount of (+)-valencene (1) that was converted to nootkatol (2 and 3) and (+)-nootkatone (4)

10 ml purification buffer (50 mM TrisHCl, pH 7.5, 500 mM NaCl, 100 μM phenylmethanesulfonyl fluoride). Cells were lysed by sonification on ice (3× 1,5 min, 1 min intermission), cell debris was removed by centrifugation (35,000 × g, 30 min, 4°C), the soluble protein fraction was recovered and filtered through a 0,2 μm filter.

Recombinant expression of PdR and Pdx was achieved by transformation of E. coli BL21(DE3) with either pET28a-camA or pET28a-camB. Transformants were selected on LB-agar plates with kanamycin (30 μg ml⁻¹). 400 ml LB with kanamycin were inoculated with 2 ml from an overnight culture – grown from a single colony – and grown at 37°C and 180 rpm, until the OD$_{600}$ reached approximately 0.7. 100 μM IPTG was added and the culture was grown for 17 h at 25°C and 160 rpm. The soluble protein fraction was recovered as described for CYP109B1.

Purification of CYP109B1, PdR and Pdx was done by immobilized metal affinity chromatography (IMAC) with a Talon® resin (7 ml bed volume). Protein lysates were applied to the column, which was pre-equilibrated with 5 column volumes of purification buffer. Non-specifically bound proteins were washed from the column with 4 column volumes of purification buffer with 5 mM imidazol, before the bound protein was eluted with purification buffer containing 100 mM imidazol. 5% Glycerol was added to the eluate and it was dialyzed two times against 2 l of 50 mM TrisHCl, pH 7.5, containing 5% glycerol, 100 μM phenylmethanesulfonyl fluoride and frozen at -20°C until use.

Determination of protein concentration

The P450 expression levels were estimated using the CO-difference spectral assay as described previously [35,36] using $\varepsilon_{450-490} = 91$ mM⁻¹ cm⁻¹.

The concentration of PdR was determined as the average of the concentration calculated from each of the three wavelength 378, 454 and 480 nm using extinction coefficients 9.7, 10.0 and 8.5 mM⁻¹ cm⁻¹ [37].

The concentration of Pdx was determined as the average concentration calculated from the two wavelength 415 and 455 nm using extinction coefficients 11.1 and 10.4 mM⁻¹ cm⁻¹ [37].

In vitro activity reconstitution of CYP109B1

A reconstituted in vitro system for conversion of (+)-valencene (1) by CYP109B1 was set up as described in [25] except for addition of catalase. The components were mixed in a 1,5 ml reaction tube, incubated at 30°C for 2 min and 200 μM NADH were added. Absorption at 340 nm was followed spectro-photometrical and NADH-consumption was calculated using $\varepsilon = 6.22$ mM⁻¹ cm⁻¹. For product identification the setup was slightly modified as follows: 1 μM PdR, 10 μM Pdx, 1 μM CYP109B1, 5 units of glucose-6-phosphate dehydrogenase, 4 mM glucose-6-phosphat and 1 mM MgCl$_2$. After incubation, the internal standard (-)-carvone (50 μM) was added, samples were extracted with 400 μl ethyl acetate and the extracts were analyzed by GC/MS.

Construction of whole-cell biocatalysts

E. coli BL21(DE3) was used as a host for the gene expression. The BL21(DE3) cells were correspondingly transformed with plasmids harboring P450-genes and grown in 50 ml of M9 expression-medium, which is based on M9 medium [32], supplemented with 1% casamino acids, 20 μg ml⁻¹ thymine, 0.1 μM FeSO$_4$, 80 μg ml⁻¹ 5-aminolevulinic acid and 50 μg ml⁻¹ carbenicillin. For protein expression the Overnight Express™ Autoinduction System 1 (Novagen, Darmstadt, Germany) was added according to the supplier's manual. The cultivation was carried out for 24 h at 25°C. Cells were collected by centrifugation (2400 × g) and resuspended in 10 ml of aqueous CV2 buffer (50 mM potassium phosphate, pH 7.5, 2% glycerol, 0.1 μM IPTG, 50 μg ml⁻¹ carbenicillin), whereupon the cww was adjusted to 70 g l⁻¹. The substrate was added to a final concentration of 2 mM for the biooxidation assay either directly or from a stock solution dissolved in DMSO to yield a final DMSO concentration in the reaction mixture of 2%. Samples were then split in 2 ml aliquots. For the biphasic systems 10 or 20% (v/v) of an organic solvent (isooctane, *n*-octane, dodecane or hexadecane) was added to the CV2 buffer. The biotransformation was carried out at 30°C under shaking at 200 rpm for 8 h. After incubation the internal standard (-)-carvone (50 µM) was added and samples were extracted with 1 ml of ethyl acetate for GC or GC/MS analysis. A negative control was performed with *E. coli* BL21(DE3) transformed with pT7NS-camAB.

For biooxidation reactions on a larger scale, 20 ml of CV2 buffer were set up with 70, 140 or 280 g_{cww} l^{-1} in round bottom flasks (70 g cww correspond to 18.4 g cdw). (+)-Valencene (1) was added from concentrated stock solutions in DMSO to achieve final concentrations of 2, 3 or 4 mM (+)-valencene (1) and 2% DMSO in the reaction mixture. The reaction was carried out at 30°C for 8 h under stirring with a magnetic stirrer. After incubation the internal standard (-)-carvone (50 µM) was added and samples were extracted with 10 ml ethyl acetate for GC analysis.

Viability assays

E. coli viability during the biotransformation was monitored by taking cell aliquots from the reaction mixtures at certain time points. 40 µl of the aqueous phase were diluted in serial dilutions, plated on Luria broth agar plates containing ampicillin (100 µg ml^{-1}) and after incubation at 30°C for 24 h the grown colonies were counted.

Product analysis

The concentrations of (+)-valencene (1) and its conversion to products was determined by GC analysis using a GC2014 (Shimadzu, Kyoto, Japan) equipped with an Equity-5 column (30 m × 0.25 mm × 0.25 µm, Supelco, Pennsylvania, USA). The injector and detector temperatures were set at 250 and 270°C, respectively. 1 µl of a sample was injected in a split mode of 4, with helium as carrier gas. The column temperature was maintained at 120°C for 4 min, ramped to 250°C at a rate of 10°C min^{-1} and held at 250°C for 5 min. For quantitative GC analysis the FID response was calibrated for (+)-valencene (1) and (+)-nootkatone (4). Mixtures of CV2 buffer containing (+)-valencene (1) or (+)-nootkatone (4) in final concentrations of 50 to 2500 µM and (-)-carvone from a 5 mM stock solution in ethanol (final concentration 50 µM) as an internal standard were extracted with 1 ml of ethyl acetate and analyzed as described. The ratio of the area of the substrate to that of the internal standard was plotted against the substrate concentration to give a straight-line calibration plot.

Mass spectra were acquired on a GC/MS-QP2010 (Shimadzu) equipped with a FS-Supreme-5 column (30 m × 0.25 mm × 0.25 µm, Chromatographie Service GmbH, Langerwehe, Germany). The same setup as for the GC-analysis was used. The products were identified by their characteristic mass fragmentation patterns by comparison with mass spectra of authentic reference compounds and by comparison with mass spectra reported elsewhere [16].

Determination of distribution coefficients

The distribution coefficients (log *D*) for (+)-valencene (1), *trans*-nootkatol (3) and (+)-nootkatone (4) were determined by dissolving increasing amounts of the respective compound (final concentration 500 to 5000 µM) in a 50% (v/v) mixture of aqueous CV2 buffer and an organic solvent (isooctane, *n*-octane, dodecane or hexadecane) in a total volume of 10 ml. After equilibration for 4 h at 25°C under vigorous shaking, phases were separated by centrifugation. 500 µl of each the organic and the aqueous phase were recovered separately and the internal standard (-)-carvone (final concentration 50 µM) was added. The aqueous phase was extracted with 1 ml of ethyl acetate and 500 µl of ethyl acetate were added to the organic phase. Both phases were analyzed quantitatively by GC. Subsequently the log *D* values were calculated.

Synthesis of cis- (2) and trans-nootkatol (3)

390 mg (+)-nootkatone (4) was dissolved in 2 ml dry diethyl ether and added dropwise under stirring at 0°C on ice to a suspension of 130 mg LiAlH$_4$ in 8 ml dry diethyl ether. After complete conversion, the solution was cooled down to -30°C and 20 ml saturated sodium potassium tartrate solution were added. The solution was allowed to warm to ambient temperature and was stirred for 16 h. The aqueous phase was extracted with diethyl ether (4 × 20 ml). The extracts were combined and dried over anhydrous MgSO$_4$. Solvents were removed at 40°C. The raw product was applied to a silica gel column (ethyl acetate:petroleum ether 1:10). Fractions were analyzed by GC/MS and NMR. Nootkatol was isolated as a mixture of *cis*- (2) and *trans*-nootkatol (3) with a ratio of *cis*:*trans* approximately 1:9. Solvents were removed from the nootkatol fraction at 40°C until no change in mass occurred. Nootkatol (2 and 3) (350 mg, 89%) was retained as colorless oil.

^{1}H-NMR (500 MHz, CDCl$_3$, for numbering see figure 1): δ = 0.89 (3 H, d, *J* 6.9, 4-Me), 0.95 (1 H, br s, 2-OH), 1.00 (3 H, s, 5-Me), 1.21 (1 H, dddd, *J* 13.9, 12.4, 12.4, 4.3, 8-H$_{ax}$), 1.37 (1 H, dd, *J* 12.7, 12.7, 10.0, 3-H$_{ax}$), 1.52 (1 H, dqd, *J* 13.0, 6.8, 2.1, 4-H), 1.71 (3 H, dd, *J* 1.3, 1.1, 13-Me), 1.77 (1 H, dddd, *J* 12.3, 2.3, 1.6, 1.3, 8-H$_{eq}$), 1.78 – 1.84 (1 H, m, 3-H$_{eq}$), 1.86 (1 H, ddd, *J* 12.8, 2.7, 2.7, 6-H$_{eq}$), 2.12 (1 H, ddd, *J* 14.1, 4.2, 2.6, 9-H$_{eq}$), 2.21 – 2.28 (1 H, m, 9-H$_{ax}$), 2.29 – 2.37 (1 H, m, 7-H), 4.23 – 4.27 (1 H, m, 2-H$_{ax}$), 4.67 – 4.70 (2 H, m, 2× 14-H$_{ax}$), 5.32 (1 H, ddd, *J* 2.6, 1.8, 1.8, 1-H$_{ax}$) ppm.

^{13}C-NMR (500 MHz, CDCl$_3$): δ = 15.4 (4-Me), 18.2 (5-Me), 20.8 (13-Me), 32.4 (C-8), 32.9 (C-6), 37.3 (C-10), 38.2 (C-5), 39.3 (C-4), 40.8 (C-7), 44.6 (C-9), 68.0 (C-2), 108.6 (C-14), 124.3 (C-1), 146.1 (C-10), 150.2 (C-13) ppm.

Competing interests
The authors declare that they have no competing interests.

Authors' contributions
MG carried out the screening of the P450 library, the biphasic experiments and drafted the manuscript. VBU has made substantial contributions to conception and design of experiments and participated in writing the manuscript. AA, KM and MI participated in construction of the P450 library and interpretation of the screening results. All authors read the manuscript and gave final approval of the version to be published.

Additional material

Additional file 1
12% SDS-PAGE of protein expression in recombinant E. coli and purified CYP109B1, putidaredoxin reductase (PdR) and putidaredoxin (Pdx). The picture provided shows protein expression as follows: Uninduced E. coli whole-cells (line 1); induced CYP109B1-, PdR- and Pdx-co-expressing strain (line 2); induced PdR- and Pdx-co-expressing strain (line 3); soluble protein fraction of CYP109B1 overexpression strain (line 4); purified CYP109B1 (line 5); purified PdR (line 6); purified Pdx (line 7); PageRuler™ Unstained Protein Ladder (line M). The molecular weights were estimated with 45.0 kDa for CYP109B1, 43.5 kDa for PdR and 11.6 kDa for Pdx.
Click here for file
[http://www.biomedcentral.com/content/supplementary/1475-2859-8-36-S1.png]

Additional file 2
Mass spectra of oxidation products derived from (+)-valencene conversion. The mass spectra provided correspond to the GC-chromatogram shown in Figure 2. Each spectrum is numbered according to the peak number given in Figure 2. The numbers represent spectra of cis-nootkatol (peak 2), trans-nootkatol (peak 3), (+)-nootkatone (peak 4) and overoxidation products (peaks 5, 6 and 7). Mass spectra of 2, 3 and 4 were compared to those of authentic reference compounds that were either commercially available or have been synthesized chemically in our lab.
Click here for file
[http://www.biomedcentral.com/content/supplementary/1475-2859-8-36-S2.png]

Acknowledgements
We wish to thank Prof. Shinobu Oda (Kanazawa Institute of Technology) for helpful comments and Sebastian Kriening (Universitaet Stuttgart) for synthesis of nootkatol. MG, VBU and RDS acknowledge the support of this work by Arbeitsgemeinschaft industrieller Forschungsvereinigungen "Otto von Guericke" e.V. (AiF), Deutsche Forschungsgemeinschaft (SFB706) and Ministerium für Wissenschaft, Forschung und Kunst Baden-Württemberg.

References
1. Fraga BM: **Natural sesquiterpenoids.** *Nat Prod Rep* 2004, **21:**669-693.
2. Chappell J: **Valencene synthase – a biochemical magician and harbinger of transgenic aromas.** *Trends Plant Sci* 2004, **9:**266-269.
3. Sharon-Asa L, Shalit M, Frydman A, Bar E, Holland D, Or E, Lavi U, Lewinsohn E, Eyal Y: **Citrus fruit flavor and aroma biosynthesis: isolation, functional characterization, and developmental regulation of Cstps1, a key gene in the production of the sesquiterpene aroma compound valencene.** *Plant Journal* 2003, **36:**664-674.
4. Fraatz MA, Berger RG, Zorn H: **Nootkatone – a biotechnological challenge.** *Appl Microbiol Biotechnol* 2009, **83:**35-41.
5. Arantes SF, Farooq A, Hanson JR: **The preparation and microbiological hydroxylation of the sesquiterpenoid nootkatone.** *J Chem Res-S* 1999:248-248A.
6. Salvador JAR, Melo ML, Neves AS: **Copper-catalysed allylic oxidation of Delta(5)-steroids by t-butyl hydroperoxide.** *Tetrahedron Lett* 1997, **38:**119-122.
7. Salvador JAR, Clark JH: **The allylic oxidation of unsaturated steroids by tert-butyl hydroperoxide using surface functionalised silica supported metal catalysts.** *Green Chem* 2002, **4:**352-356.
8. Hashimoto T, Asakawa Y, Noma Y, Murakami C, Tanaka M, Kanisawa T, Emura M: **Nootkatone manufacture with Chlorella valencene or hydroxyvalencene.** *Japanese patent* 2003. No. 2003070492
9. Furusawa M, Hashimoto T, Noma Y, Asakawa Y: **Biotransformation of citrus aromatics nootkatone and valencene by microorganisms.** *Chem Pharm Bull (Tokyo)* 2005, **53:**1423-1429.
10. Furusawa M, Hashimoto T, Noma Y, Asakawa Y: **Highly efficient production of nootkatone, the grapefruit aroma from valencene, by biotransformation.** *Chem Pharm Bull (Tokyo)* 2005, **53:**1513-1514.
11. Hashimoto T, Asakawa Y, Noma Y, Murakami C, Furusawa M, Kanisawa T, Emura M, Mitsuhashi K: **Manufacture of nootkatone with Mucor sp.** *Japanese patent* 2003. No. 2003250591
12. Kaspera R, Krings U, Nanzad T, Berger RG: **Bioconversion of (+)-valencene in submerged cultures of the ascomycete Chaetomium globosum.** *Appl Microbiol Biotechnol* 2005, **67:**477-483.
13. de Kraker JW, Schurink M, Franssen MCR, Konig WA, de Groot A, Bouwmeester HJ: **Hydroxylation of sesquiterpenes by enzymes from chicory (Cichorium intybus L.) roots.** *Tetrahedron* 2003, **59:**409-418.
14. Willershausen H, Graf H: **Mikrobieller Abbau von Lignin.** *CLB Chemie in Labor und Biotechnik* 1996, **47:**24-28.
15. Huang R, Christenson P, Labuda IM: **Production of natural flavours by laccase catalysts.** *European patent* 2001. No. 1083233
16. Takahashi S, Yeo YS, Zhao Y, O'Maille PE, Greenhagen BT, Noel JP, Coates RM, Chappell J: **Functional characterization of premnaspirodiene oxygenase, a cytochrome P450 catalyzing regio- and stereo-specific hydroxylations of diverse sesquiterpene substrates.** *J Biol Chem* 2007, **282:**31744-31754.
17. Bernhardt R: **Cytochromes P450 as versatile biocatalysts.** *J Biotechnol* 2006, **124:**128-145.
18. Peterson JA, Lu JY, Geisselsoder J, Grahamlorence S, Carmona C, Witney F, Lorence MC: **Cytochrome-P-450terp – Isolation and Purification of the Protein and Cloning and Sequencing of Its Operon.** *Journal of Biological Chemistry* 1992, **267:**14193-14203.
19. Bell SG, Sowden RJ, Wong LL: **Engineering the haem monooxygenase cytochrome P450cam for monoterpene oxidation.** *Chemical Communications* 2001:635-636.
20. Celik A, Flitsch SL, Turner NJ: **Efficient terpene hydroxylation catalysts based upon P450 enzymes derived from actinomycetes.** *Org Biomol Chem* 2005, **3:**2930-2934.
21. Dietrich M, Eiben S, Asta C, Do TA, Pleiss J, Urlacher VB: **Cloning, expression and characterisation of CYP102A7, a self-sufficient P450 monooxygenase from Bacillus licheniformis.** *Appl Microbiol Biotechnol* 2008, **79:**931-940.
22. Cryle MJ, Stok JE, De Voss JJ: **Reactions catalyzed by bacterial cytochromes P450.** *Aust J Chem* 2003, **56:**749-762.
23. Isin EM, Guengerich FP: **Complex reactions catalyzed by cytochrome P450 enzymes.** *Biochim Biophys Acta* 2007, **1770:**314-329.

24. Tassaneeyakul W, Guo LQ, Fukuda K, Ohta T, Yamazoe Y: **Inhibition selectivity of grapefruit juice components on human cytochromes P450.** *Arch Biochem Biophys* 2000, **378:**356-363.
25. Sowden RJ, Yasmin S, Rees NH, Bell SG, Wong LL: **Biotransformation of the sesquiterpene (+)-valencene by cytochrome P450cam and P450BM-3.** *Org Biomol Chem* 2005, **3:**57-64.
26. Agematu H, Matsumoto N, Fujii Y, Kabumoto H, Doi S, Machida K, Ishikawa J, Arisawa A: **Hydroxylation of testosterone by bacterial cytochromes P450 using the Escherichia coli expression system.** *Biosci Biotechnol Biochem* 2006, **70:**307-311.
27. Yasutake Y, Imoto N, Fujii Y, Fujii T, Arisawa A, Tamura T: **Crystal structure of cytochrome P450 MoxA from Nonomuraea recticatena (CYP105).** *Biochem Biophys Res Commun* 2007, **361:**876-882.
28. Buhler B, Schmid A: **Process implementation aspects for biocatalytic hydrocarbon oxyfunctionalization.** *J Biotechnol* 2004, **113:**183-210.
29. Witholt B, de Smet MJ, Kingma J, van Beilen JB, Kok M, Lageveen RG, Eggink G: **Bioconversions of aliphatic compounds by Pseudomonas oleovorans in multiphase bioreactors: background and economic potential.** *Trends Biotechnol* 1990, **8:**46-52.
30. Wubbolts MG, Favre-Bulle O, Witholt B: **Biosynthesis of synthons in two-liquid-phase media.** *Biotechnol Bioeng* 1996, **52:**301-308.
31. Leon R, Fernandes P, Pinheiro HM, Cabral JM: **Whole-cell biocatalysis in organic media.** *Enzyme Microb Tech* 1998, **23:**483-500.
32. Sambrook J, Russell D: *Molecular Cloning: A Laboratory Manual* Third edition. New York: Cold Spring Harbor Laboratory Press; 2001.
33. Arisawa A, Agematu H: **A Modular Approach to Biotransformation Using Microbial Cytochrome P450 Monooxygenases.** In *Modern Biooxidation* First edition. Edited by: Schmid RD, Urlacher VB. Weinheim: Wiley-VCH; 2007:177-192.
34. Kunst F, Ogasawara N, Moszer I, Albertini AM, Alloni G, Azevedo V, Bertero MG, Bessieres P, Bolotin A, Borchert S, *et al.*: **The complete genome sequence of the gram-positive bacterium Bacillus subtilis.** *Nature* 1997, **390:**249-256.
35. Omura T, Sato R: **The Carbon Monoxide-Binding Pigment of Liver Microsomes. II. Solubilization, Purification, and Properties.** *J Biol Chem* 1964, **239:**2379-2385.
36. Omura T, Sato R: **The Carbon Monoxide-Binding Pigment of Liver Microsomes. I. Evidence for Its Hemoprotein Nature.** *J Biol Chem* 1964, **239:**2370-2378.
37. Purdy MM, Koo LS, Ortiz de Montellano PR, Klinman JP: **Steady-state kinetic investigation of cytochrome P450cam: interaction with redox partners and reaction with molecular oxygen.** *Biochemistry* 2004, **43:**271-281.

2.1.2.1. Supplementary Material

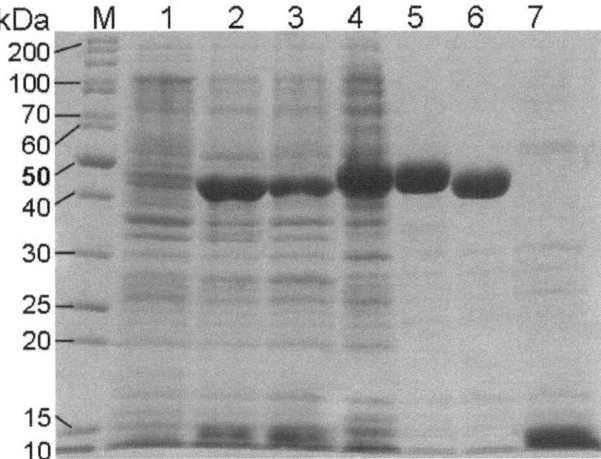

Figure 2-9 (Additional file 1): 12% SDS-PAGE of protein expression in recombinant *E. coli* and purified CYP109B1, putidaredoxin reductase (PdR) and putidaredoxin (Pdx). The picture provided shows protein expression as follows: Uninduced *E. coli* whole-cells (line 1); induced CYP109B1-, PdR- and Pdx-co-expressing strain (line 2); induced PdR- and Pdx-co-expressing strain (line 3); soluble protein fraction of CYP109B1 overexpression strain (line 4); purified CYP109B1 (line 5); purified PdR (line 6); purified Pdx (line 7); PageRuler™ Unstained Protein Ladder (line M). The molecular weights were estimated with 45.0 kDa for CYP109B1, 43.5 kDa for PdR and 11.6 kDa for Pdx.

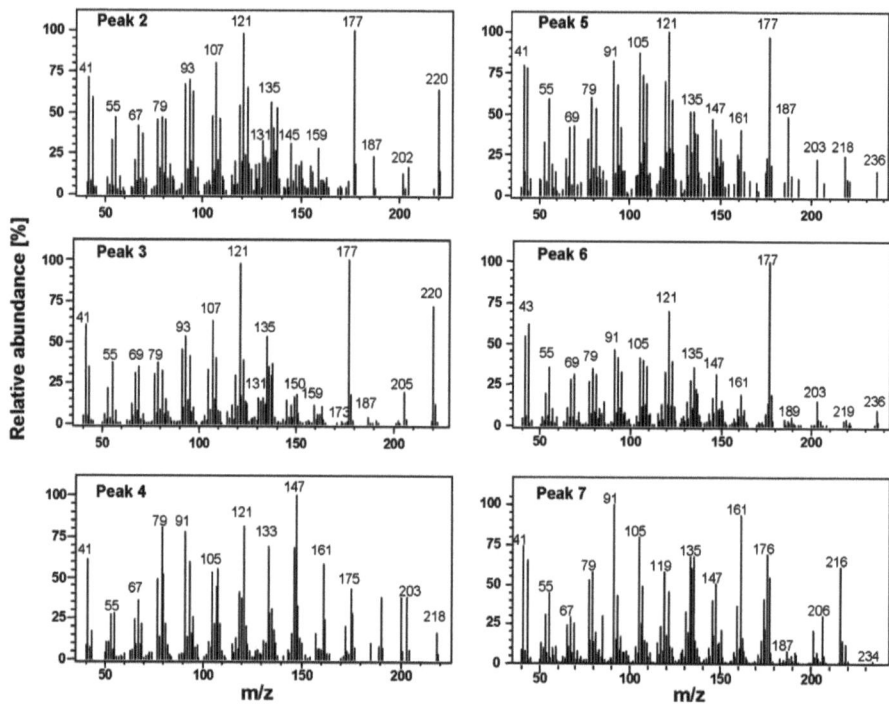

Figure 2-10 (Additional file 2): Mass spectra of oxidation products derived from (+)-valencene conversion. The mass spectra provided correspond to the GC-chromatogram shown in Figure 2. Each spectrum is numbered according to the peak number given in Figure 2. The numbers represent spectra of cis-nootkatol (peak 2), trans-nootkatol (peak 3), (+)-nootkatone (peak 4) and overoxidation products (peaks 5, 6 and 7). Mass spectra of 2, 3 and 4 were compared to those of authentic reference compounds that were either commercially available or have been synthesized chemically in our lab.

2.1.3. Manuscript: Regioselective hydroxylation of norisoprenoids by CYP109D1 from *Sorangium cellulosum*

Material from this chapter appears in:

Yogan Khatri[a], Marco Girhard[b], Anna Romankiewicz[c], Michael Ringle[a], Frank Hannemann[a], Vlada B. Urlacher[b], Michael C. Hutter[d] and Rita Bernhardt[a,*], 2010, Regioselective hydroxylation of norisoprenoids by CYP109D1 from *Sorangium cellulosum*, *Applied Microbiology and Biotechnology*, 88(2):485-495.

[a] Department of Biochemistry, Saarland University, 66041 Saarbrücken, Germany
[b] Institute of Biochemistry, Heinrich-Heine-University Düsseldorf, 40225 Düsseldorf, Germany
[c] Institute of Technical Biochemistry, Universität Stuttgart, 70569 Stuttgart, Germany
[d] Center for Bioinformatics, Saarland University, 66041, Saarbrücken, Germany
* Corresponding author

Material is reprinted by permission of Springer; the original manuscript is available online at: http://www.springerlink.com

2.1.3.1. Abstract

Sesquiterpenes are particularly interesting as flavorings and fragrances or as pharmaceuticals. Regio- or stereo-selective functionalizations of terpenes are one of the main goals of synthetic organic chemistry, which are possible through radical-reactions but are not selective enough to introduce desired chiral alcohol function into those compounds. Cytochrome P450 monooxygenases are versatile biocatalysts and are capable of performing selective oxidations of organic molecules. We were able to demonstrate that CYP109D1 from *Sorangium cellulosum* So ce56 functions as a biocatalyst for the highly regio-selective hydroxylation of norisoprenoids, α- and β-ionone, which are important aroma compounds of floral scents. The substrates α- and β-ionone were regio-selectively hydroxylated to 3-hydroxy-α-ionone and 4-hydroxy-β-ionone, respectively, which was confirmed by ^{1}H NMR and ^{13}C NMR. The results of docking α-ionone and β-ionone into a homology model of CYP109D1 gave a rational explanation for the regio-selectivity of the hydroxylation. Kinetic studies revealed that α- and β-ionone can be hydroxylated with nearly identical V_{max} and K_m values. This is the first comprehensive investigation of the regio-selective hydroxylation of norisoprenoids by CYP109D1.

2.1.3.2. Introduction

Terpenes and terpenoids are the most diverse family of natural products serving a range of important physiological and societal functions (Roberts 2007). Over 40,000 different terpenoids have been isolated, mainly from plants, but also from animal and microbial species (Withers and Keasling 2007). Although the oxygenated products of terpenes are highly desirable compounds, their selective oxidation under mild conditions is still a great challenge. Regio- and stereo-selective functionalizations of terpenes, especially those of terpene hydrocarbons, is known to be a difficult task for organic chemistry, and the few approaches described in the literature show several disadvantages with respect to reaction complexity and specificity (Stanislaw et al. 2002; Duetz et al. 2003). Therefore, application of cytochrome P450 monooxygenases (P450 or CYP) can be useful (Bernhardt 2006). It was already shown that P450s are capable of mediating selective oxidization of different terpenes and terpenoids, often with a high activity (Girhard et al. 2009, 2010; Urlacher et al. 2006).

Ionones are norisoprenoids that are substantial aroma components of floral scents (Winterhalter and Rouseff 2002), and thus attract the attention of flavor and fragrance industry. Due to their organoleptic properties and the distinctive fine rose scents, manufacturers of perfumes, soaps, cosmetics, and fine chemicals are particularly interested in those compounds (Pybus 1999). α-ionone shows a more floral smell than β-ionone. Besides this, ionones are also appreciated as synthetic building blocks (Buchecker et al. 1973; Colombo et al. 1992). Among them, 4-hydroxy-β-ionone is an important intermediate for the synthesis of carotenoids (Eschenmoser et al. 1981; Brenna et al. 2002) and also of deoxyabscisic acid, a synthetic analogue of the phytohormone abscisic acid (Larroche et al. 1995). It is also used in vitamin A (retinol) production for cosmetics and toiletries. Moreover, α-ionone and its derivatives have been described as an effective attractant for fruit fly (*Bactrocera latifrons*) and they form important components of insect lures, which can favor insect pollination (Ishda et al. 2008). Only few methods of preparing hydroxyl-derivatives of ionones, such as the chemical optical resolution of α-ionone (Haag et al. 1980) or the selective enzymatic hydrolysis of 4-hydroxy-β-ionone, have been published (Kakeya et al. 1991). Microbial transformation of α- and β-ionone has been reported using several fungal strains, especially *Aspergillus* sp. (Larroche et al. 1995). Likewise, the conversion of β-ionone into 4-hydroxy-β-ionone and the conversion of α-ionone into 3-hydroxy-α-ionone by several *Streptomyces* sp. has been studied (Mikami et al. 1981; Yamazaki et al 1988; Lutz-Wahl et al. 1998). The biotransformation of α-ionone and β-ionone catalyzed by P450s derived from actinomycetes (Celik et al. 2005) and the engineered cytochrome P450$_{BM3}$ have also been reported (Appel et al. 2001). However, in all these cases, besides the 3-hydroxylation of α-ionone, a mixture of different products has been observed. In the case of β-ionone as a substrate, the wild type P450$_{BM3}$ and its mutants were able to produce only 4-hydroxy-β-ionone (Urlacher et al. 2006), whereas CYP101B1 and CYP101C1 from *Novosphingobium aromaticivorans* did not show regio-selectivity of the hydroxylation resulting in both 3-hydroxy-β-ionone and 4-hydroxy-β-ionone products (Bell et al. 2010).

Here, we describe the function of CYP109D1 from myxobacterium *Sorangium cellulosum* So ce56. Myxobacteria are a group of gram-negative, mainly soil-dwelling bacteria which belong to the delta subdivision of proteobacteria and show a complex life cycle involving multicellular development, cellular differentiation and fruiting bodies formation (McCudry 1989; Reichenbach 2004). The ~13.1 Mb genome of the model strain *Sorangium cellulosum* So ce56 is the largest yet discovered in bacteria (Schneiker et al. 2007). Varieties of biosynthetic gene clusters involved in natural product formation as well as numerous complex and unusual biosynthetic processes within *Sorangium cellulosum* have been identified (Wenzel and Müller, 2007; 2009). The selected strain So ce56 produces the natural secondary metabolites chivosazol, etnangien and myxochelin (Schneiker et al., 2007; Wenzel and Müller, 2007; Menche et al., 2008). We were able to show that CYP109D1 is a regio-selective hydroxylase of norisoprenoids. We demonstrated the exclusive production of 3-hydroxy-α-ionone from α-ionone and of 4-hydroxy-β-ionone from β-ionone using a reconstituted P450 system.

2.1.3.3. Material and methods

Materials

All chemicals and reagents were of highest available grade. The expression vector pET17b and the expression host *E. coli* BL21(DE3) were purchased from Invitrogen and Novagen, respectively. The α- and β-ionone were of GC grade and purchased from Fluka. The alkane standard solution (C8-C20) was purchased from Sigma-Aldrich.

Protein expression and purification

Genomic DNA from *S. cellulosum* So ce56 was isolated as described before (Perlova et al. 2006). The DNA fragment encoding CYP109D1 (CAN94796) was prepared by the polymerase chain reaction using genomic DNA of *S. cellulosum* So ce56 as a template. The PCR primers were designed to introduce a *Nde*I restriction site at the 5'-end of the fragment and a *Hind*III restriction site with 6-histidine tag at the 3'-end, respectively. Primers used in the reaction were as follows: Forward: 5'-CCAAT<u>CAT*ATG*</u>GAAACCGAGACCGCCCCGAGCCC–3'. The letters underlined indicate an engineered restriction site of *Nde*I, including the initiation codon ATG (in italic). Reverse: 5'-AATTG<u>GAAGCTT</u>*TCA***GTGATGGTGATGGTGATG**GGCGGTGGCGCGG CTC–3'. The underlined letters show a site for *Hind*III and the stop codon TCA (in italic) is also incorporated. The bold letters in the reverse primer indicate the hexa-histidine (His$_6$)-tag. The CYP109D1 PCR product was cloned in the expression vector pET17b with *Nde*I and *Hind*III restriction sites, and coexpressed with the molecular chaperones GroES/GroEL. Briefly, for the coexpression, host *E. coli* BL21(DE3), was transformed with two expression plasmids, pET17b_CYP109D1 and pGro12_GroES/GroEL (Nishihara et al. 1998; Zöller et al. 2008), and cultured overnight in terrific broth (TB) containing 100 µg ml^{-1} ampicillin and 50 µg ml^{-1} kanamycin. The overnight culture was one hundred-fold diluted in 200 ml of the TB medium in 2 l baffled flasks containing 100 µg ml^{-1} ampicillin and 50 µg ml^{-1} kanamycin. The culture was grown at 37°C and agitated at 95 rpm until the optical density at 600 nm reached 1.2. During culturing, the heme synthesis was supported by the addition of 0.8 mM of delta-aminolevulinic acid (δ-ALA) while the expression was induced with 1 mM isopropyl β-D-1-thiogalactopyranoside (IPTG). To induce the chaperone expression 4 mg ml^{-1} final concentration of arabinose was added. At the same time 50 µg ml^{-1} of ampicillin was also added to overcome ampicillin degradation at high temperatures or its deactivation by beta-lactamases. The cultures were grown at 28°C for an additional 24-36 h shaking at 95 rpm. The cultures were harvested by centrifugation at 4500 rpm and the pellet was stored at -20°C until purification of the protein. The cell pellet was disrupted by sonication and the soluble His$_6$-tagged CYP109D1 was purified by immobilized metal ion affinity chromatography (IMAC) using TALON™ (Clontech) resin. Purification was carried out following the manufacturer's instructions. Fractions containing the heterologously expressed protein were collected and concentrated for size exclusion chromatography, which was performed by using a Superdex 75 (GE Healthcare) column and 10 mM potassium phosphate buffer, pH 7.5, with a constant flow rate of 0.1 ml min^{-1}. Suitable fractions were collected, concentrated, and stored at -20°C. The active form of CYP109D1 was confirmed by CO-difference spectrum measurements (Omura and Sato, 1964).

The mammalian truncated adrenodoxin Adx $_{(4-108)}$ (Uhlmann et al. 1994) and adrenodoxin reductase (AdR) were expressed and purified as described before (Sagara et al. 1993).

UV-visible absorption spectroscopy

UV-visible spectra for CYP109D1 were measured at room temperature on a double-beam spectrophotometer (UV-2101PC, SHIMADZU, Japan). CYP109D1 (5 µM) in buffer A (10 mM potassium phosphate buffer, pH 7.5 containing 20% glycerol) was used for the spectral measurements of the oxidized and reduced form. CYP109D1 was reduced by the addition of a small amount of sodium dithionite. The P450 was converted to its high-spin substrate bound form by the addition of α- or β-ionones, to a final concentration of 5 µM for both the substrates.

The concentration of CYP109D1 was estimated using CO-difference spectra assuming $\varepsilon_{(450-490)}$ = 91 mM^{-1}cm^{-1} according to the method as described before (Omura and Sato 1964). Briefly, the solution of P450 enzyme in 10 mM potassium phosphate buffer (pH 7.5) containing 20% glycerol was reduced with few grains of sodium dithionite. It was split into two cuvettes and a base line was recorded between 400 and 500 nm. The sample cuvette was bubbled gently with carbon monoxide for 1 min and a spectrum was recorded.

Substrate-induced spin-state shift

Spin-state shifts upon substrate binding were assayed at 25°C under aerobic conditions using an UV–visible scanning photometer (UV-2101PC, Shimadzu, Japan) equipped with two tandem quartz cuvettes (Hellma, Müllheim, Germany). One chamber of each cuvette contained 3 µM CYP109D1 in 800 µl of 10 mM potassium phosphate buffer, pH 7.5, whereas the second chamber contained buffer alone. The titration of the substrates (α- and β-ionone dissolved in DMSO) was done by adding small (< 2 µl) aliquots of an appropriate stock of the substrate into the P450 containing chamber of the sample cuvette. An equal amount of the substrate was also added into the buffer containing chamber of the reference cuvette, and spectral changes between 360 and 500 nm were recorded. After being saturated with the substrate, the substrate dissociation constant (K_D) for CYP109D1 with the substrate was calculated by fitting the peak-to-trough difference against substrate concentration to a non linear tight binding quadratic equation (Williams and Morrison 1979), which accounts the quantity of the substrate consumed in complex with the P450 in determining the K_D value for the substrate. The relevant equation (Eqn. 2) is:

$$\Delta A = \frac{A_{max} \times (K_d + [E] + [S]) - ((K_d + [E] + [S])^2 - 4 \times [E] \times [S])^{1/2}}{2[E]} \quad \text{Eqn. 2}$$

ΔA represents the observed peak peak-to-trough absorbance difference at each substrate addition, A_{max} is the maximum absorbance difference at substrate saturation, [E] is the total enzyme (CYP109D1) concentration and [S] is the substrate concentration. The data fitting was performed using Origin 8.1 software. All titrations were done for three times and the K_D values reported are the mean for the three sets of the experimental data.

In vitro assay

An *in vitro* conversion of the norisoprenoids (α- and β-ionone) with the reconstituted heterologous electron partners, Adx and AdR, was done with CYP109D1 to investigate the functional properties of the enzyme. The reactions were performed in a final volume of 0.5

ml consisting of 10 mM potassium phosphate buffer, pH 7.4 containing 20% glycerol, 200 µM substrate and a NADPH-regenerating system consisting of glucose-6-phosphate (5 mM), glucose-6-phosphate dehydrogenase (1 unit) and magnesium chloride (1 mM). Reconstitution with the redox partners Adx and AdR was performed according to the ratio of CYP109D1:Adx:AdR of 1:10:1 (1 µM : 10 µM : 1 µM). The reaction was started with an addition of NADPH (500 µM) and the mixture was incubated for 30 min in a thermomixture (Thermomixture, eppendorf). The reaction was stopped and extracted twice with an equal volume of ethyl acetate. The organic phases were pooled and the volume was reduced to 200 µl with a flow of nitrogen gas. The samples were run on either thin layer Silica Gel 60 F254 plates and/or directly analyzed using gas chromatography coupled with mass spectrometry or HPLC as mentioned below.

Thin layer chromatography (TLC)

Thin layer chromatography (TLC) was used to monitor the conversion of α- and β-ionone. Samples were spotted onto TLC aluminum sheets (5 x 10 cm^2, silica gel layer thickness, 0.2 mm; Silica Gel 60 F254; Merck), and the plates were developed in a solvent tank containing n-hexane : ethyl acetate (3 : 2). Products were visualized firstly with UV light at 254 nm and then confirmed by spraying with a vanillin solution (2.5% (w/v) in ethanol:sulfuric acid (H_2SO_4) (95:5) and subsequent heating at 110°C for 2 min.

Gas chromatography coupled with mass spectrometry (GC/MS) analysis

GC/MS was performed on a Shimadzu GC/MS QP2010 (Kyoto, Japan) equipped with an FS Supreme-5 column (0.25 mm x 30 m, 0.25 µm, CS-Chromatography Service, Langerwehe, Germany). Helium was used as a carrier gas with a total flow rate of 6.8 ml min^{-1} (column flow was 0.63 ml min^{-1}). The following program was used for the analysis of α- and β-ionone, and their products. The column temperature was controlled at 150°C for 1 min. The temperature was then raised to 250°C at 20°C min^{-1}. It was kept at 250°C for 5 min. The temperatures of injector and the interface were fixed at 250°C and 1 µl of a sample was injected with a split of 5.

The Kovats retention indices for the products of α- and β-ionone were calculated from a homologue of the standard straight-chain alkanes (C8-C20), which was within the linear temperature programming area of the ionone conversion. The retention indices (RI) for the temperature programmed chromatography were calculated by equation 3:

$$RI_{(TC)} = \frac{RT_{(TC)} - RT_{(n)}}{RT_{(N)} - RT_{(n)}} \times (100 \times Z) + (100 \times n) \qquad \text{Eqn. 3}$$

(TC) is the name of the target compound, (n) is the smaller alkane directly eluting before (TC), (N)' is the larger alkane directly eluting after (TC), Z is the difference of the number of carbon atoms in the smaller and larger alkane, RT is the retention time and RI is the retention index (Kovats 1958).

Isolation of 3-hydroxy-α-ionone from the reaction mixture

For the identification of the product of α-ionone oxidation, the *in vitro* reaction described above was scaled up 120 times. The organic phase was dried using vacuum conditions. The obtained product was completely dissolved in a mixture of n-hexane : ethyl acetate (3 : 2) and purified on a 150 ml silica gel column. The complete separation of remaining α-ionone and the product was monitored by TLC. The purified product fractions were pooled and dried in a rotary vacuum device (SpeedVac Concentrator 5301, Eppendorf). From the residue, 10 mg were used for NMR analysis (^{1}H and ^{13}C NMR) (Advance 500, Bruker Biospin GmbH, Rheinstetten, Germany).

Kinetic analysis of α- and β-ionone conversion

For the estimation of the kinetic constants, increasing concentrations of the ionones (0-150 µM) with fixed concentrations of CYP109D1 (1 µM), Adx (10 µM) and AdR (1 µM) were used. All other components and methods for the assay were as described above.

After evaporation of the organic (ethyl acetate) phase, the substrates (α- and β-ionone) were resuspended in acetonitrile:water (40:60) and separated on a Jasco reversed phase HPLC system (Tokyo, Japan) composed of an auto sampler AS-2050 plus, pump PU-2080, gradient mixer LG-2080-02 and an UV-detector UV-2075. A reversed phase column (Nucleodur R100-5 C18ec, particle size 3 µm, length 125 mm and internal diameter 4 mm, Macherey-Nagel) was used to separate the substrates and the products. The substrates were monitored at 240 nm and the column temperature was kept constant at 25°C with a peltier oven. The mobile phase was a mixture of acetonitirile : water (40 : 60) for both the substrates at a flow of 1 ml min^{-1}. 10 µl of the samples were injected for analysis. The peaks were identified by using the ChromPass software (V.1.7.403.1, Jasco).

V_{max} and K_m values were determined by plotting the product formation rate versus the corresponding substrate concentration using a hyperbolic fit (Michaelis-Menten kinetics) applying SIGMAPLOT (Systat Software, San Jose, CA, USA).

Computational methods

P450eryF (CYP107A1) from *Saccharopolyspora erythreae* (accession code Q00441) shows a sequence identity of 31.6% to CYP109D1 of *Sorangium cellulosum* So ce56. The corresponding crystallographic structure (pdb entry 1Z8P) was chosen as structural template. The alignment generated with CLUSTALW (version 2) (Larkin et al. 2007) was used as input for SWISS-MODEL (Arnold et al. 2007; Schwede et al. 2003; Guex and Peitsch 1997). The coordinates of the heme-porphyrin atoms from the template structure were added subsequently to the obtained homology model. The vicinity around the porphyrin was structurally relaxed by force field optimization using the AMBER3 set of parameters as implemented in HYPERCHEM version 6.02. The carboxylate groups of the heme form ionic hydrogen-bonds with the side chains of Arg293, Arg104, His100, and His349 in CYP109D1. Compounds for docking were generated manually and energetically optimized using the MM+ force field as implemented in HYPERCHEM. AUTODOCK (version 4.00) was applied for docking of α- and β-ionone into the homology model of CYP109D1 (Huey et al. 2007; Morris et al. 2007). The Windows version 1.5.2 of Autodock Tools was used to compute Gasteiger-Marsili charges for the enzyme and the ligands (Sanner 1999). A partial charge of +0.400 e was assigned manually to the heme-iron, which corresponds to Fe(II), that was compensated by adjusting the partial charges of the ligating nitrogen atoms to −0.348 e. Flexible bonds of the ligands were assigned automatically and verified by manual inspection. A cubic grid box (50 x 50 x 50 points with a grid spacing of 0.375 Å) was centered above the heme-iron at the putative position of the ferryl oxygen. For each of the ligands 100 docking runs were carried out applying the Lamarckian genetic algorithm using default parameter settings, except for the mutation rate that was increased to 0.05.

2.1.3.4. Results

Expression and purification of recombinant CYP109D1

CYP109D1 from *Sorangium cellulosum* So ce56 was first expressed in *E. coli* BL21 using the bacterial expression vector pCWori$^+$. However, the expression level determined by CO-difference spectral assay was only 205 nmol l^{-1} *E. coli* culture. Therefore, the CYP109D1-encoding gene was cloned in pET17b. Coexpression of pET17b_CYP109D1 was carried out with the molecular chaperones GroES/GroEL. A four-fold increase in the expression level of CYP109D1 up to 900 nmol l^{-1} of soluble enzyme was obtained. The CO-difference spectrum demonstrated a peak maximum at 450 nm.

Spectrophotometric characterization

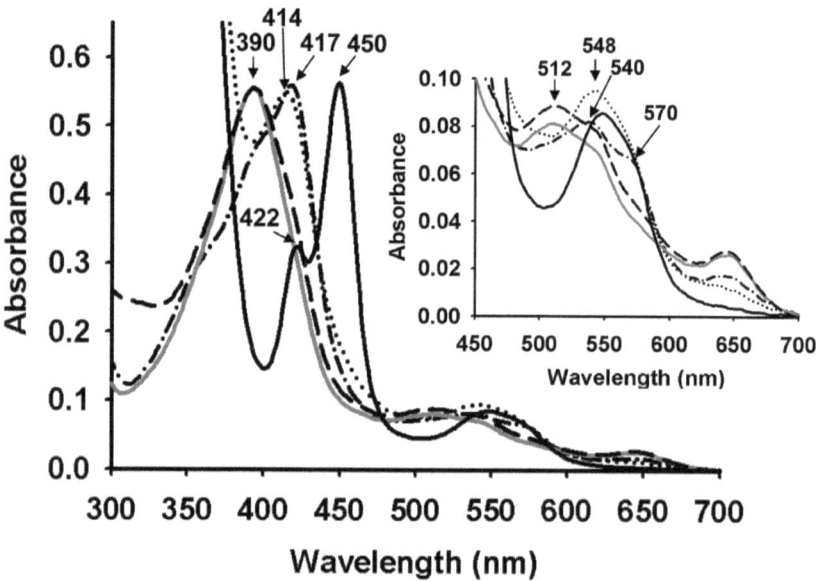

Figure 2-11: UV-visible spectral characterization of CYP109D1. UV-visible absorption spectrum of oxidized (dashed with dot), dithionite reduced (dotted line), reduced CO-complexed (solid line), α-ionone bound (grey line) and β-ionone bound (long dashed) CYP109D1. CYP109D1 (5 µM) in 10 mM potassium phosphate buffer (pH 7.5) containing 20% glycerol was used. Concentrations of both substrates were 5 µM. The inset shows a magnification of the spectrum in the α/β band region.

UV-visible absorption spectroscopy provides the primary technique for the characterization of P450 enzymes. The oxidized form of substrate free CYP109D1 of So ce56 demonstrated a large component high spin heme iron with a major Soret (γ) band at 417 nm and the smaller α and β bands at 570 nm and 540 nm, respectively (Figure 2-11). It was observed that the Soret band of CYP109D1 was split between low spin and high spin components with a Soret maximum at 417 nm. A similar spectrum was also obtained with the mutant A233G of CYP121 from *Mycobacterium tuberculosis* (McLean et al. 2008).

The dithionite reduced spectrum of CYP109D1 showed a slightly diminished absorption maximum of the Soret band at 414 nm (Figure 2-11). Under aerobic condition, it seemed difficult to reduce CYP109D1 completely, which could be due to the fast re-oxidation of the heme iron as compared to the rate of the reduction of the ferric heme by dithionite. The carbon monoxide bound form gave a typical peak at 450 nm with a small shoulder at 422 nm. The former is indicative of native enzyme (the cytochrome P450 species), but the latter indicates that a proportion of the enzyme has lost the cysteinate ligation following dithionite reduction and exposure to CO. The binding of both the substrates, α- and β-ionone, shifted the Soret (417 nm) maxima to ~390 nm (Figure 2-11).

Substrate binding spectra

Binding of the substrates α- and β-ionone to CYP109D1 induced a shift in the equilibrium of the heme iron spin towards the high spin form leading to changes in the Soret region (type I spectrum) which is characterized by a peak at ~390 and a trough at ~427 nm (Figure 2-12). Titration of the enzyme with α- and β-ionone revealed K_D values of 6.22 ± 0.39 µM (Figure 2-12a) and 6.76 ± 0.84 µM (Figure 2-12b), respectively. Both the ionones showed tight binding with CYP109D1 with almost similar K_D values.

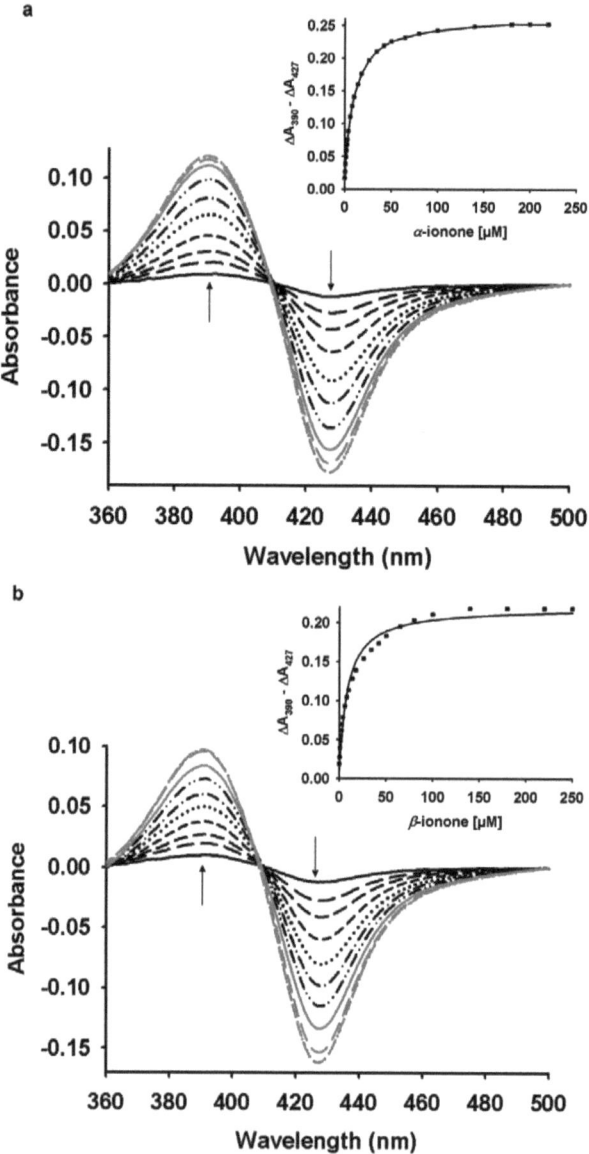

Figure 2-12: Spectral shifts induced by binding of α- and β-ionone to CYP109D1. Varying concentrations of α-ionone **(a)** and β-ionone **(b)**, dissolved in DMSO, were titrated to a solution of CYP109D1 (3 µM) in 10 mM potassium phosphate buffer (pH 7.5) containing 20% glycerol, with the starting spectrum subtracted from the subsequent traces. The final concentrations of the substrate were 0.250 µM - 250 µM. Arrows represent the direction of the peak on increasing

substrate concentrations. The lines in solid black, long dashed, medium dashed, short dashed, dotted, dashed with single dot, dashed with double dots, solid grey, long dashed grey, short dashed grey and dotted grey represent substrate concentrations of 0.250, 1, 2, 4, 8, 14, 26, 50, 100, 220 and 250 µM, respectively. The inset shows absorbance changes plotted against the respective concentrations of α-ionone (a) and β-ionone (b) used for the titration of CYP109D1. The points of each titrations were fitted to the quadratic equation cited in the Method section and the values for K_D for α-ionone and β-ionone were calculated to be 6.22 ± 0.39 µM and 6.76 ± 0.84 µM, respectively, with a regression coefficient (R) of 0.99 and 0.96, respectively.

Conversion of α-ionone

In general, cytochrome P450 monooxygenases require electron transfer partners for activity (Bernhardt 2006; Hannemann et al. 2007). Therefore, the heterologous redox partners adrenodoxin (Adx) and adrenodoxin reductase (AdR), which have been used already to successfully reconstitute the activity of other bacterial P450s, namely CYP106A2 and CYP109B1 (Virus and Bernhardt 2008; Lisurek et al. 2008; Hannemann et al. 2006; Girhard et al. 2010), were added to our system. GC/MS analyses demonstrated that CYP109D1 was able to convert α-ionone into a single product. The product was obtained at a retention time of 6.88 min (Figure 2-13). The retention index (RI), the retention time normalized to the adjacently eluting n-alkanes, for the product was calculated in reference with the obtained alkane pattern (supplementary material, Figure 2-17a). The calculated RI for the product of α-ionone was 1645 (supplementary material, Figure 2-17b). A similarity search of its mass spectrum within the NIST arbitrary library did not lead to an unambiguous identification of the product. Therefore, we purified and characterized this product.

Identification of the product of α-ionone oxidation

A large scale conversion of α-ionone and a purification of the product on a silica gel column were performed to obtain sufficient amounts of the reaction product. 16 mg of the purified product were obtained after silica gel chromatography. The identification of the presence of a functional -OH group in the product was identified using Infra-red (IR) spectroscopy, and was confirmed by ^{1}H and ^{13}C NMR spectroscopy. The product of α-ionone was found to be 3-hydroxy-α-ionone (Figure 2-13). In the TLC, the values of the retention factor (R_f) for the α-ionone and its product were 0.67 and 0.30, respectively (data not shown).

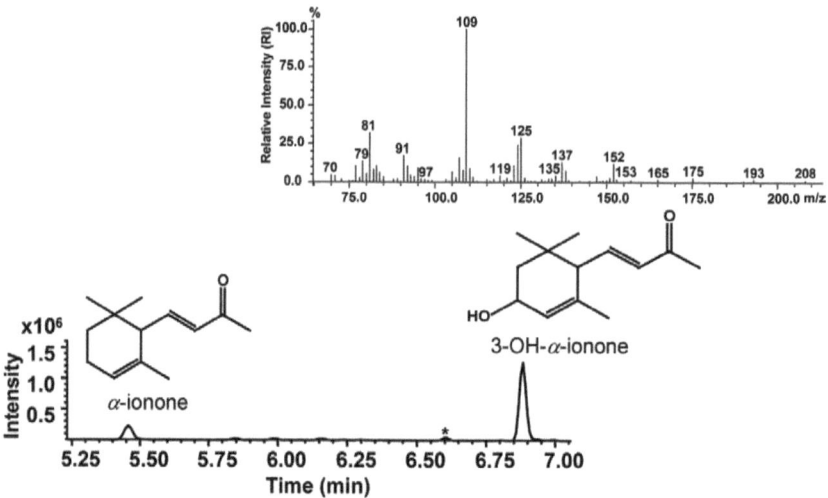

Figure 2-13: GC diagram of α-ionone conversion in the presence of CYP109D1. The chromatogram represents the conversion of α-ionone. The product peak (t_R = 6.88 min) is identified as 3-hydroxy-α-ionone. The structures of both the substrate and the product are shown. The '*' represents a non-specific product. The inset shows the MS spectrum of 3-hydroxy-α-ionone.

The detailed results of GC/MS, IR and NMR spectroscopy for the α-ionone product were as follows:

GC/MS (*EI*): (t_R = 6.88 min) (RI 1645), (208, M+, 82%), (109, 100%), (81, 32.8%), (125, 29.4%), (124, 24.8%), (91, 17.9%), (107, 16.3%), (79, 13.5%), (137, 13.1%), (152, 12.4%), (123, 11.3%), (83, 10.2%), (138, 7.3%), (175, 2.8%), (193, 2.0%), (165, 1.7%). **IR:** v_{max}/cm^{-1} (FT). 3449

^{1}H NMR: δ_H (CDCl$_3$; 200 MHz), 6.47 (1 H, dd, *J* = 15.7 and 10 Hz); 6.03 (1 H; d, *J* = 15.8 Hz); 5.56 (1 H, br); 4.20 (1 H, br); 2.43 (1 H; d, *J* = 10 Hz); 2.19 (3 H, s); 1.77 (1 H, dd, J = 13.6 and 6.0 Hz); 1.51 (3 H, s); 1.33 (1 H; dd, *J* = 13.6 and 6.6 Hz); 0.96 (3 H, s); 0.82 (3 H, s).

^{13}C NMR δ: 33.35 (C-1), 43.83 (C-2), 65.48 (C-3), 125.83 (C-4), 135.20 (C-5), 54.30 (C-6), 147.06 (C-7), 133.62 (C-8), 198.30 (C-9), 27.20 (C-10), 29.70 (C-11), 24.21 (C-12), 22.66 (C-13).

Conversion of β-ionone

CYP109D1 was also able to convert β-ionone into a single product. The product was obtained at a retention time of 7.07 min (Figure 2-14). The calculated retention index (RI) for the product was 1675 (supplementary material, Figure 2-17c). The similarity search of the mass spectra within the NIST mass spectra library showed that the MS-data was 90% similar to that of 4-hydroxy-β-ionone. This product was further compared with authentic 4-hydroxy-β-ionone (Urlacher et al. 2006; Haag et al. 1980) and finally confirmed as 4-hydroxy-β-ionone. In the TLC, the values of the retention factor (R_f) for the β-ionone and its product were 0.93 and 0.34, respectively (data not shown).

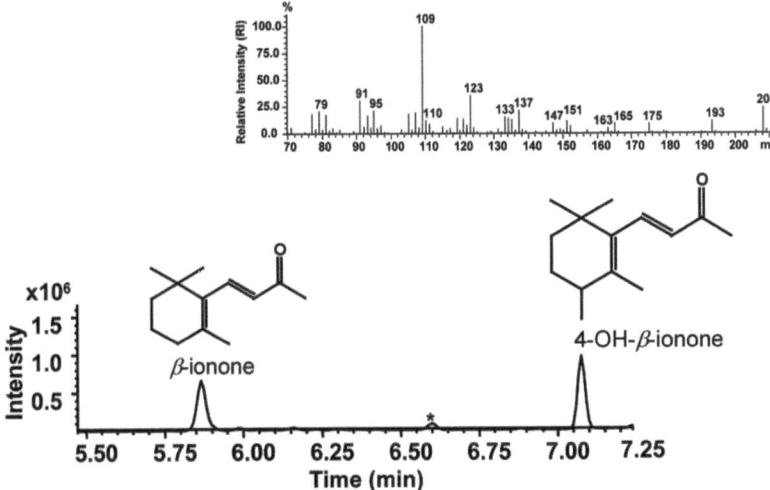

Figure 2-14: GC diagram of β-ionone conversion in the presence of CYP109D1. The chromatogram represents the conversion of β-ionone. The product peak (t_R = 7.07 min) is identified as 4-hydroxy-β-ionone. The structure of both the substrate and product are shown. The '*' represents a non-specific product. The inset shows the MS spectrum of 4-hydroxy-β-ionone

The detailed results of GC/MS for the β-ionone product were as follows:

GC/MS (*EI*): (t_R = 7.07 min) (RI 1675), (208, M+, 46%), (109, 100%), (123, 30.3%), (91, 23.8%), (208, 21.4%), (137, 18.6%), (95, 18.5%), (79, 15.5%), (107, 15.4%), (105, 14.6%), (119, 11.8%), (151, 11.2%), (193, 8.4%), (165, 7.9%), (175, 6.5%), (163, 4.4%).

Kinetic analysis of CYP109D1

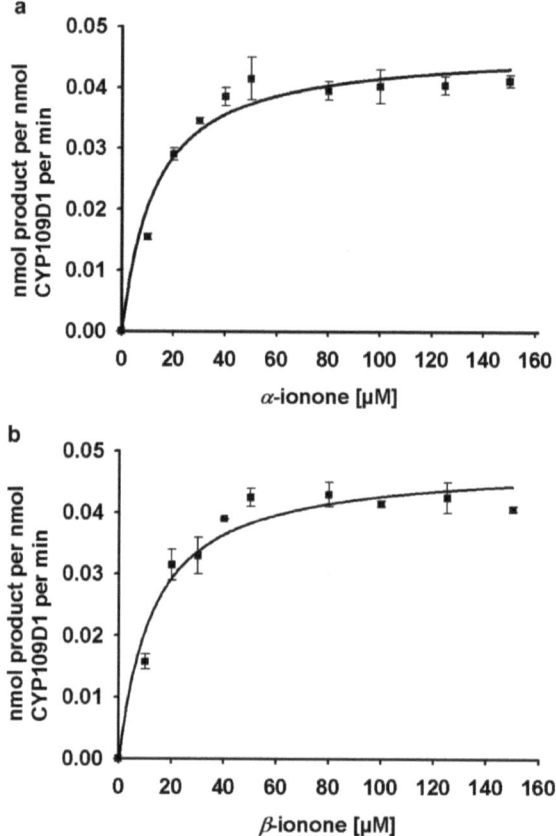

Figure 2-15: Determination of kinetic parameters for α-ionone (**a**) and β-ionone (**b**) conversions catalyzed by the CYP109D1. The activity was reconstituted with the heterologous redox partners, Adx and AdR. The ratio of CYP109D1:Adx:AdR was 1:10:1. The hydroxylations of the ionones were analyzed by HPLC as described in Method section. Bars on each point represent the standard deviation of three individual experiments. Different concentrations of α- and β-ionone (0-150 µM) were used as substrates. Hyperbolic fits of the 3-hydroxy-α-ionone product of α-ionone (**a**) and the 4-hydroxy-β-ionone product of β-ionone (**b**) are shown.

After the functionality of the enzyme was demonstrated by performing CYP109D1-dependent substrate conversion assays using α- and β-ionone as substrates, HPLC analysis was done for the kinetic analyses. The kinetic studies were done in reconstituted enzymatic systems employing the heterologous redox partners Adx and AdR as electron transfer partners. CYP109D1 hydroxylated α-ionone to 3-hydroxy-α-ionone with a V_{max} value of 0.048 ± 0.003 nmol product per nmol CYP109D1 per min and a K_m value of 12.12 ± 3.42 µM (R = 0.99) (Figure 2-15a). β-ionone was hydroxylated to gain 4-hydroxy-β-ionone with a V_{max} of 0.049 ± 0.003 nmol product per nmol CYP109D1 per min and a K_m of 12.44 ± 3.50 µM (R = 0.99) (Figure 2-15b). The comparison of the V_{max} values for 3-hydroxy-α-ionone and 4-hydroxy-β-ionone formation from the substrates, α- and β-ionone, respectively, by CYP109D1 showed a similar rate of product formation.

Homology modeling and docking of CYP109D1 with α- and β-ionone.

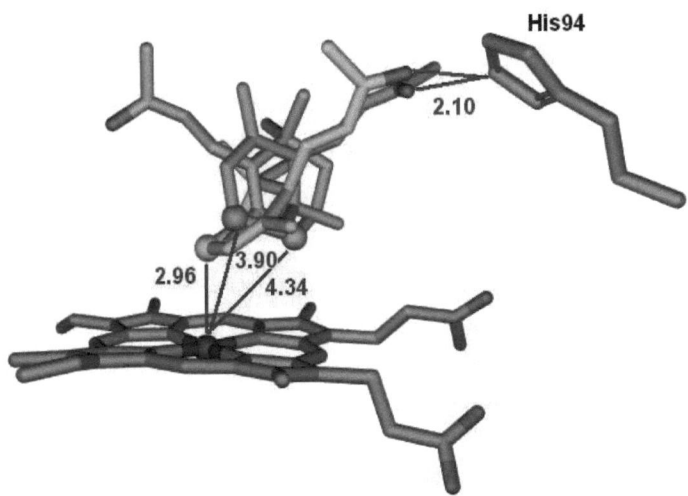

Figure 2-16: Docking positions in the homology model of CYP109D1. The *R*-isomer of α-ionone is shown in pink, the *S*-isomer in cyan, and β-ionone in orange. The respective carbonyl oxygen atoms are shown in red. The positions where hydroxylation occurs are marked as balls. Distances are given in Angstrom

The crystal structure of CYP109D1 has not been solved so far. Therefore, computer modeling of CYP109D1 and its complexes with α- and β-ionone was performed to get more insight into the structural basis for the regio-specific hydroxylation of these substrates. CYP109D1 from So ce56 has the highest amino acid identity (31.6%) with

P450eryF (CYP107A1) from *Saccharopolyspora erythreae* among the known crystal structures of bacterial cytochrome P450s deposited in the protein data bank (PDB). The alignment is shown in the supplementary material (Figure 2-18). The corresponding crystallographic structure (pdb entry 1Z8P) was chosen as structural template (see method sections for details). The carboxylate groups of the heme-moiety form ionic hydrogen-bonds with the side chains of Arg293, Arg104, His100, and His349. Both the *R*- and *S*-isomer of α-ionone were found in binding positions that have the 3-position in suitable arrangement for hydroxylation (Figure 2-16).

The distances towards the iron atom are 3.90 Å for the *R*-isomer and 2.96 Å for the *S*-isomer, respectively. Both isomers show a hydrogen-bond (2.1 Å) between the carbonyl oxygen and the NE2-nitrogen atom of His94. All other obtained binding poses exhibit alternative sites for hydroxylation much more distant. This indicates structural implications as the reason for the regioselective hydroxylation of α-ionone. Likewise, the closest distance found between the iron atom and the 4-position in β-ionone is 4.34 Å, whereas all other hydroxylation sites within the ring are again further away from the iron atom.

2.1.3.5. Discussion

Most terpenes and terpenoids are commercially valuable products with applications depending on characteristic functional groups. From the viewpoint of chemical synthesis, sesquiterpenes are particularly interesting as flavorings and fragrances or as pharmaceuticals. Several attempts have been undertaken to discover new biologically active terpenoids with even higher market value. However, the selective oxy-functionalization of terpenes and oxidation of the resulting alcohols to the corresponding carbonyls are still challenging to organic chemistry due to low specificities and the need of hazardous and expensive catalysts and reactants (Duetz et al. 2003; Lochynski et al. 2002).

An increasingly attractive alternative to both, chemical synthesis and conventional isolation methods is the enzymatic or microbial production of terpene and terpenoid derivatives from natural raw material, which has been the target of intensive research activities during the past decades (Buehler and Schmidt 2004). Nowadays, biocatalysis offers promising synthetic alternatives due to the huge natural diversity of oxygenating and oxidizing enzymes acting as highly regio- and stereo-selective catalysts. The great majority of these reactions, including some terpene oxidations, were conducted with whole cells expressing cytochrome P450 monooxygenases which, however, resulted in rather low productivities and the formation of by-products (de Carvalho and da Fonseca 2006; Duetz et al. 2001). Although P450s usually show low activity, instability and dependence on redox partners as well as the costly cofactors NAD(P)H, these enzymes still possess a great technical potential because of their high regio- and stereo-selectivity (Duetz et al. 2003; Bernhardt 2006; Urlacher and Smith 2002).

Within this study we cloned and expressed the cytochrome P450 monooxygenase CYP109D1 from *Sorangium cellulosum* So ce56 which is able to perform the conversion of norisoprenoids in a reconstituted system with Adx and AdR as redox partners. Though there are 8 ferredoxins and 3 ferredoxin reductases available in So ce56 (Ewen et al. 2009), only two of the ferredoxins (Fdx2 and Fdx8) in combination with one reductase (FdR_B) were able to sustain the activity of CYP109D1-dependent conversion of fatty acids, however, the efficiency of the conversion with heterologous redox partners, Adx and AdR, was much higher than that with the native redox partners (Khatri et al. unpublished data). Furthermore, we have also shown that the conversion of myristic acid by CYP109B1 from *Bacillus subtilis*, the other member of the CYP109 family, was also more efficient with Adx and AdR compared with other redox partners (Girhard et al. 2010). Therefore, the heterologous redox partners, Adx and AdR, were used as redox partners for the conversion of α- and β-ionone. CYP109D1 catalyzed the selective hydroxylation of α- and β-ionone, which are valuable fragrance constituents. We were able to show that CYP109D1 can convert α-ionone exclusively into 3-hydroxy-α-ionone, which was confirmed by ^{1}H and ^{13}C NMR. We further demonstrated that CYP109D1 converted β-ionone exclusively into 4-hydroxy-β-ionone. This was confirmed by the comparison to an authentic standard (Urlacher et al. 2006; Haag et al, 1980). The bioconversion of α-ionone and β-ionone to their corresponding mono-hydroxylated derivatives has also been examined before using a recombinant *Escherichia coli* whole cell system expressing cytochrome P450 SU1 and SU2, or P450 SOY (Celik et al. 2005). However, in this case the hydroxylation was shown to be non-specific revealing several side products. Likewise, P450$_{BM3}$ of *Bacillus megaterium* also showed conversion of α-ionone into mixtures of different hydroxylated and/or epoxidated products (Carmichael and Wong 2001). Even the triple P450$_{BM3}$ mutant (F78Q L188Q A74G) produced a mixture of different hydroxylated products from α-ionone (Appel et al. 2001). Along with this, the P450$_{BM3}$ mutant F87V also produced four products from α-ionone (Urlacher et al. 2006). Therefore, it is of special interest that CYP109D1, despite being a non-improved wild type enzyme, was able to give a single hydroxylated product (3-hydroxy-α-ionone) from α-ionone. The hydroxylation of β-ionone into 4-hydroxy-β-ionone as described here was also published for the wild type of P450$_{BM3}$ and its improved mutants (e.g. A74E F87V P386S) (Urlacher et al. 2006). The chemical basis for the regio-selectivity of hydroxylation of α-ionone and its regioisomer β-ionone by CYP109D1 and other P450s is not completely understood, but the stereo chemistry implies that the electronic activation of the allylic carbon at C-4 by the C-5=C-6 double bond governs the regio-selectivity of β-ionone hydroxylation, whereas in the isomeric α-ionone, C-3 in allylic position to the double bond at C-4=C-5 is also most susceptible to an oxidative attack by CYP109D1 and other P450s.

To further elucidate the structural basis for the regio-selectivity of the CYP109D1-catalyzed hydroxylation, we performed docking studies using a homology model of CYP109D1. According to the computed interaction energies, both the α- and β-ionone should bind in the range of micro-molar affinity (with respect to the uncertainty of the underlying scoring function). The obtained docking positions showed both ionones in corresponding orientations enabling the experimentally observed regio-selective hydroxylation (Figure 2-16). Other binding positions that would allow alternative hydroxylation in the 4-position of α-ionone (both R- and S-isomer) were not found. Likewise, no plausible docking positions allowing the hydroxylation in the 3-position of β-ionone were obtained.

In summary, our results demonstrate that CYP109D1 acts as a highly regioselective hydroxylase for the oxidation of norisoprenoids, α- and β-ionone. Thus, CYP109D1 could be a potential candidate for the selective oxidation of terpenes and terpenoids, which can generate valuable compounds for flavor and fragrance industries. A similar whole-cell system as created for CYP106A2 coexpressing the P450 as well as the redox partners Adx and AdR (Hannemann et al. 2006), could be used for the bioconversion of terpenes and terpenoids with potential biotechnological applications.

2.1.3.6. Acknowledgments

This work was supported by the fellowship awarded by Deutscher Akademischer Austausch Dienst (DAAD) to Y. K., Deutsche Bundesstiftung Umwelt (DBU) to M. R. and a grant of the Fonds der Chemischen Industrie to R. B. We are thankful to Wolfgang Reinle for expression and purification of Adx and AdR. V.U., M.G. and A.R. thank the German Research Foundation (DFG, SFB706) for financial support.

2.1.3.7. References (to chapter 2.1.3)

Appel D, Lutz-Wahl S, Fischer P, Schwaneberg U, Schmid, RD (2001) A P450$_{BM3}$ mutant hydroxylates alkanes, cycloalkanes, arenes and heteroarenes. J Biotechnol 88:167–171.

Arnold K, Bordoli L, Kopp J, Schwede T (2006) The SWISS-MODEL Workspace: A web-based environment for protein structure homology modeling. Bioinformatics 22:195-201.

Bell SG, Dale A, Rees NH, Wong LL (2010) A cytochrome P450 class I electron transfer system from *Novosphingobium aromaticivorans*. Appl Microbiol Biotechnol 86:163-75.

Bernhardt R (2006) Cytochromes P450 as versatile biocatalysts. J Biotechnol 124:128–145.

Brenna E, Fuganti C, Serra S, Kraft, P (2002) Optically active ionones and derivatives: Preparation and olfactory properties. Eur J Org Chem 967-978.

Buchecker R, Egli R, Regel-Wild H, Tscharner C, Eugster CH, Uhde G, Ohloff G (1973) Absolute konfiguration der enantiomeren α-cyclogeraniumsäuren, α-cyclogeraniale, α-ionone, γ-ionone, α- und ε-Carotine. Helv Chim Acta 56:2548-2563.

Buehler B, Schmid A (2004) Process implementation aspects for biocatalytic hydrocarbon oxyfunctionalization. J Biotechnol 113:183–210.

Carmichael AB, Wong LL (2001). Protein engineering of *Bacillus megaterium* CYP102: the oxidation of polycyclic aromatic hydrocarbons. Eur J Biochem 268:3117–3125.

Celik A, Flitsch SL, Turner NJ (2005) Efficient terpene hydroxylation catalysts based upon P450 enzymes derived from actinomycetes. Org Biomol Chem 3:2930-2934.

Colombo MI, Zinczuk J, Ruveda EA (1992) Synthetic routes to forskolin. Tetrahedron. 48:963-1037.

de Carvalho CC, da Fonseca MM (2006) Biotransformation of terpenes. Biotechnol Adv 24:134–142.

Duetz WA, Bouwmeester H, van Beilen JB, Witholt B (2003) Biotransformation of limonene by bacteria, fungi, yeast, and plants. Appl Microbiol Biotechnol 61:269–277.

Duetz WA, van Beilen JB, Witholt B (2001) Using proteins in their natural environment: potential and limitations of microbial whole-cell hydroxylations in applied biocatalysis. Curr Opin Biotechnol 12:419-425.

Eschenmoser W, Uevelhart P, Eugster, CH (1981) Synthesis and structure of the enantiomeric 6-hydroxy-α-ionone and cis- and trans-5,6-dihydroxy-5,6-dihydro-β-ionone. Helv Chim Acta 64:2681–2690.

Ewen KM, Hannemann F, Khatri Y, Perlova O, Kappl R, Krug D, Hüttermann J, Müller R, Bernhardt R (2009) Genome mining in *Sorangium cellulosum* So ce56: identification and characterization of the homologous electron transfer proteins of a myxobacterial cytochrome P450. J Biol Chem 284:28590-8.

Girhard M, Klaus T, Khatri Y, Bernhardt R, Urlacher VB (2010) Characterization of the versatile monooxygenase CYP109B1 from *Bacillus subtilis*. Appl Microbiol Biotechnol 87:595-607. doi: 10.1007/s00253-010-2472-z.

Girhard M, Machida K, Itoh M, Schmid RD, Arisawa A, Urlacher VB (2009) Regioselective biooxidation of (+)-valencene by recombinant *E. coli* expressing CYP109B1 from *Bacillus subtilis* in a two-liquid-phase system. Microb Cell Fact 8:36.

Guex N, Peitsch MC (1997) SWISS-MODEL and the Swiss-PdbViewer: An environment for comparative protein modelling. Electrophoresis 18:2714-2723.

Haag A, Eschenmoser W, Eugster CH (1980) Synthese von (−)-(R)-4-hydroxy-β-ionone and (−)-(5R,6S)-5-Hydroxy-4,5-dihydro-α-ionone aus (−)-(S)-α-ionone. Helv Chim Acta 63:10–5.

Hannemann F, Bichet A, Ewen KM, Bernhardt R (2007) Cytochrome P450 systems—biological variations of electron transport chains. Biochim Biophys Acta 1770:330–344.

Hannemann F, Virus C, Bernhardt R (2006) Design of an *Escherichia coli* system for whole cell mediated steroid synthesis and molecular evolution of steroid hydroxylases. J Biotechnol 124:172-181.

Huey R, Morris GM, Olson AJ, Goodsell DS (2007) A semiempirical free energy force field with charge-based desolvation. J Comput Chem 28:1145-1152.

Ishida T, Enomoto H, Nishida R (2008) New attractants for males of the solanaceous fruit fly *Bactrocera latifrons*. J Chem Ecol 34:1532-1535.

Kakeya H, Sugi T, Ohta H (1991) Biochemical preparation of optically active 4-hydroxy-β-ionone and its transformation of (S)-6-hydroxy-α-ionone. Agric Biol Chem 55:1873-1876.

Kovats E (1958) Gas-chromatographische Charakterisierung organischer Verbindungen. Teil 1: Retentionsindices aliphatischer Halogenide, Alkohole, Aldehyde und Ketone. Helv Chim Acta 41: 1915–32.

Larkin MA, Blackshields G, Brown NP, Chenna R, McGettigan PA, McWilliam H, Valentin F, Wallace IM, Wilm A, Lopez R, Thompson JD, Gibson TJ, Higgins DG (2007) CLUSTALW, version 2. Bioinformatics 23:2947-2948.

Larroche C, Creuly C, Gros JB (1995) Fed-batch biotransformation of β-ionone by *Aspergillus niger*. Appl Microbiol Biotechnol 43:222-227.

Lisurek M, Simgen B, Antes I, Bernhardt R (2008) Theoretical and experimental evaluation of a CYP106A2 low homology model and production of mutants with changed activity and selectivity of hydroxylation. Chembiochem 9:1439–1449.

Lochynski S, Kowalska K, Wawrzenczyk C (2002) Synthesis and odour characteristics of new derivatives from the carane system. Flavour Frag J 3:181–186.

Lutz-Wahl S, Fischer P, Schmidt-Dannert C, Wohlleben W, Hauer B, Schmid RD (1998) Stereo- and regioselective hydroxylation of α-Ionone by *Streptomyces* Strains. Appl Environ Microbiol 64:3878-3881.

McCudry H (1989) Fruiting Gliding Bacteria. In: Holt J (ed), Bergey's Manual of Systematic Bacteriology, Williams & Wilkins, Baltimore, pp 2139-2143.

McLean KJ, Carroll P, Lewis DG, Dunford AJ, Seward HE, Neeli R, Cheesman MR, Marsollier L, Douglas P, Smith WE, Rosenkrands I, Cole ST, Leys D, Parish T, Munro AW (2008) Characterization of active site structure in CYP121. A cytochrome P450 essential for viability of Mycobacterium tuberculosis H37Rv. J Biol Chem 283:33406-16.

Menche D, Arikan F, Perlova O, Horstmann N, Ahlbrecht W, Wenzel SC, Jansen R, Irschik H, Müllar R (2008) Stereochemical determination and complex biosynthetic assembly of ethangien, a highly potent RNA-polymerase inhibitor from the myxobacterium *Sorangium cellulosum*. J Am Chem Soc 130:14234-43.

Mikami Y, Fukunaga Y, Arita M, Kisaki T (1981) Microbial transformation of β-ionone and β-methylionone. Appl Environ Microbiol 41:610-617.

Morris GM., Goodsell DS., Halliday RS, Huey R, Hart WE, Belew RK, Olson AJ (1998) Automated docking using a lamarckian genetic algorithm and empirical binding free energy function. J Comput Chem 19:1639-1662.

Nishihara K, Kanemori M, Kitagawa M, Yanagi H, Yura T (1998) Chaperone Coexpression Plasmids:Differential and synergistic roles of DnaK-DnaJ-GrpE and GroEL-GroES in assisting folding of an allergen of Japanese cedar pollen, Cryj2, in *Escherichia coli*. Appl Environ Microbiol 64:1694-1699.

Omura T, Sato R (1964) The carbon monoxide-binding pigment of liver microsomes. I. Evidence for its hemoprotein nature. J Biol Chem 239:2370–2378.

Perlova O, Gerth K, Kaiser O, Hans A, Müller R (2006) Identification and analysis of the chivosazol biosynthetic gene cluster from the myxobacterial model strain *Sorangium cellulosum* So ce56. J Biotechnol 121:174-191.

Pybus DH, Sell CS (1999) The Chemistry of Fragrances. The Royal Society of Chemistry, London.

Reichenbach H (2004) The Myxococcales. In: Garrity GM (ed) Bergey`s Manual of Systematic Bacteriology, Springer-Verlag, New York, pp 1059-1143.

Roberts SC (2007) Production and engineering of terpenoids in plant cell culture. Nat Chem Biol 3:387-395.

Sagara Y, Wada A, Takata Y, Waterman MR, Sekimizu K, Horiuchi T (1993) Direct expression of adrenodoxin reductase in *Escherichia coli* and the functional characterization. Biol Pharm Bull 16:627–630.

Sanner MF (1999) Python: A programming language for software integration and development. J Mol Graph Model 17:57-61.

Schneiker S, Perlova O, Kaiser O, Gerth K, Alici A, Altmeyer MO, Bartels D, Bekel T, Beyer S, Bode E, Bode HB, Bolten CJ et al (2007) Complete genome sequence of the myxobacterium *Sorangium cellulosum*. Nat Biotech 25:1281-1289.

Schwede T, Kopp J, Guex N, Peitsch, MC (2003) SWISS-MODEL: an automated protein homology-modeling server. Nucleic Acids Res 31:3381-3385.

Stanislaw L, Kowaiska K, Wawrzwnczyk C (2002) Synthesis and odour characteristics of new derivatives from the carane system. Flavour Frag J 17:181-186.

Uhlmann H, Kraft R, Bernhardt R (1994) C-terminal region of adrenodoxin affects its structural integrity and determines differences in its electron transfer function to cytochrome P450. J Biol Chem 269:22557–22564.

Urlacher VB, Makhsumkhanov A, Schmid RD (2006) Biotransformation of β-ionone by engineered cytochrome P450$_{BM3}$. Appl Microbiol Biotechnol 70:53-59.

Urlacher VB, Schmid RD (2002) Biotransformations using prokaryotic P450 monooxygenases. Curr Opin Biotechnol 13:557–564.

Virus C, Bernhardt R (2008) Molecular evolution of a steroid hydroxylating cytochrome P450 using a versatile steroid detection system for screening. Lipids 43:1133–1141.

Wenzel SC, Müller R (2007) Myxobacterial natural product assembly lines: fascinating examples of curious biochemistry. Nat Prod Rep 24:1211-24.

Wenzel SC, Müller R (2009) Myxobacteria-'microbial factories' for the production of bioactive secondary metabolites. Mol Biosyst 5:567-74.

Williams JW, Morrison JF (1979) The kinetics of reversible tight-binding inhibition. Methods Enzymol 63:437-467.

Winterhalter P, Rouseff RL (2002) Carotenoid-derived aroma compounds. Washington DC, Am Chem Soc.

Withers ST, Keasling JD (2007) Biosynthesis and engineering of isoprenoid small molecules. Appl Microbiol Biotechnol 73:980-990.

Yamazaki Y, Hayashi Y, Arita M, Hieda T, Mikami, Y (1988) Microbial conversionof α-ionone, α-methylionone, and α-isomethylionone. Appl Environ Microbiol 54:2354-2360.

Zöllner A, Kagawa N, Waterman MR, Nonaka Y, Takio K, Shiro Y, Hannemann F, Bernhardt R (2008) Purification and functional characterization of human 11-β-hydroxylase expressed in *Escherichia coli*. FEBS J, 275:799-810.

2.1.3.8. Supplementary Material

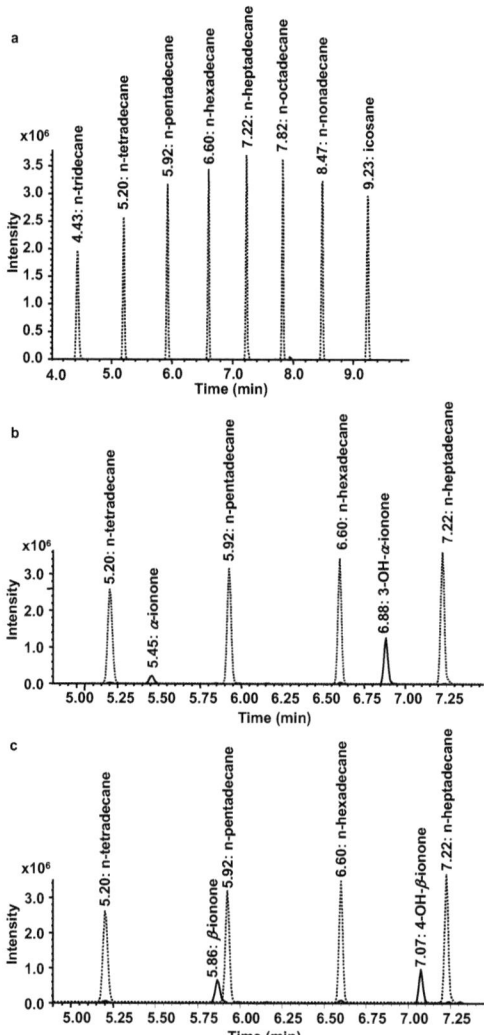

Figure 2-17: **(a)** GC diagram of alkane pattern. Standard n-alkane mixture was measured as described in the 'Methods' and the corresponding alkane peaks (C13 - C20) are shown. **(b)** The overlay of GC chromatogram of α-ionone conversion (solid line) and n-alkane (dotted line). **(c)** GC diagram of β-ionone conversion (solid line) and n-alkane (dotted line). The required parameters of the adjacent n-alkanes (n-hexadecane and n-heptadecane) of the product peak (α- or β-ionone) were used to calculate the RI value as described in the 'Methods'.

2.1 CYP109B1 from Bacillus subtilis and CYP109D1 from Sorangium cellulosum | 95

```
TARGET    7              PSPS PEQIDLSAPS VIADPYPAYR ALRGRSPVLY ARVPAGGAAG
1z8pA     2              ttvp dles----ds fhvdwyrtya elretapvtp vr-flgqdaw

TARGET                        hhhhh hhhhh ssss sss      ssss
1z8pA                         hhhhh hhhhh ssss ss s     ssss


TARGET    51   LGEPI-RAYA LLRHAEVLAA LR------D PQTFSSNVTD KIRVLPRITL
1z8pA     41   lvtgydeaka alsdlrlssd pkkkypgvev efpaylgfpe dvrnyfatnm

TARGET         sss        hh  hhh    sss                 h  hhhhhh
1z8pA          sss hhhhhh hhh        sss                 h  hhhhhh


TARGET    93   LHDDPPRHTH LRRLVSRSFT PRRIAELEPW IGRLAASLLE ATGDGP-SDL
1z8pA     91   gtsdppthtr lrklvsqeft vrrveamrpr veqitaelld evgdsgvvdi

TARGET         hhhh hhhhhh     hhhhhhhhhh hhhhhhhhhh h        sss
1z8pA          hhhh hhhhhh     hhhhh   hh hhhhhhhhhh h        sssss


TARGET    142  MGAYAMPLPM MVIATLLGIP AERYVQFRSW SESVMSYSGI PAEERASRGK
1z8pA     141  vdrfahplpi kvicellgvd ekyrgefgrw sseilvmdpe raeqrgqaar

TARGET         hh    hhhhh hhhhhhh       hhh hhhh       hhhhhhhhh
1z8pA          hh    hhhhh hhhhhhh       hhh hhhh       hhhhhhhhh


TARGET    192  AMVDFFAAEL EARRRAPSGD LISALVEAEI DG-ARLDTPE AVGFCVGLLV
1z8pA     191  evvnfildlv errrtepgdd llsalirvqd dddgrlsade ltsialvlll

TARGET         hhhhhhhhhh hhhhh      hhhhhhh    hhh hhhhhhhhh
1z8pA          hhhhhhhhhh hhhhh      hhhhhhh    hhh hhhhhhhhh


TARGET    241  AGNDTTTNLI GNMAHLLSER PELYRRAQQD RSLVGPIIEE TLRHSSPVQR
1z8pA     241  agfessvsli gigtyllth  pdqlalvrrd psalpnavee ilryiappet

TARGET         h  hhhhhh hhhhhhh   hhhhhhhh    hhhhhh hhhh   ss
1z8pA          h  hhhhhh hhhhhhh   hhhhhhhh    hhhhhh hhhh   ss


TARGET    291  LLRVTTRPVD VSGVMIPAGH LVDVVFGAAN RDPAVFEEPD AFRLDRPPAE
1z8pA     291  ttrfaaeeve iggvaipqys tvlvangaan rdpkqfpdph rfdvtrdtrg

TARGET         ssssss  ss  s  sss     s ssssssshhh
1z8pA          ssssss  ss  s  sss     s ssssssshhh


TARGET    341  HLAFGQGTHF CIGAALARME ARIALNALLD CYESITPGEA -PPLRQTRAI
1z8pA     341  hlsfgqgihf cmgrplakle gevalralfg rfpalslgid addvvwrrsl

TARGET                    hhhhhh hhhhhhhhhh h     sss
1z8pA                     hhhhhh hhhhhhhhhh h     sss


TARGET    390  MPLGFESLPL VLRR
1z8pA     391  llrgidhlpv rldg-

TARGET         sss s
1z8pA          sss ss
```

Figure 2-18: Alignment and secondary structure of CYP109D1 (target) and the template (pdb entry 1z8p). α-helices are denoted by 'h' whereas β-sheets are indicated by 's'.

2.2. CYP152A2 from *Clostridium acetobutylicum*

2.2.1. Manuscript: Cytochrome P450 monooxygenase from *Clostridium acetobutylicum*: A new α-fatty acid hydroxylase

Material from this chapter appears in:

Marco Girhard[a], Stefanie Schuster[b], Matthias Dietrich[a], Peter Dürre[b] and Vlada B. Urlacher[a], 2007, Cytochrome P450 monooxygenase from *Clostridium acetobutylicum*: A new α-fatty acid hydroxylase, *Biochemical and Biophysical Research Communications*, 362(1):114-119.

[a] Institute of Technical Biochemistry, Universität Stuttgart, 70569 Stuttgart, Germany
[b] Institute of Microbiology and Biotechnology, University of Ulm, 89069 Ulm, Germany
* Corresponding author

Material is reprinted by permission of Elsevier; the original manuscript is available online at: http://www.sciencedirect.com/science/journal/0006291X

Cytochrome P450 monooxygenase from *Clostridium acetobutylicum*: A new α-fatty acid hydroxylase

Marco Girhard [a], Stefanie Schuster [b], Matthias Dietrich [a], Peter Dürre [b], Vlada B. Urlacher [a,*]

[a] *Institute of Technical Biochemistry, University of Stuttgart, 70569 Stuttgart, Germany*
[b] *Institute of Microbiology and Biotechnology, University of Ulm, 89069 Ulm, Germany*

Received 19 July 2007
Available online 7 August 2007

Abstract

Cytochrome P450 monooxygenase from the anaerobic microorganism *Clostridium acetobutylicum* (CYP152A2) has been produced in *Escherichia coli*. CYP152A2 was shown to bind a broad range of saturated and unsaturated fatty acids and corresponding methyl esters and demonstrated a high peroxygenase activity of up to 200 min^{-1} with myristic acid. Although a high concentration of hydrogen peroxide of 200 μM was necessary for high activities of the enzyme, it led to a fast enzyme inactivation within 2–4 min. This might reflect the natural function of CYP152A2 as a rapid hydrogen peroxide scavenging enzyme. In two different reconstituted systems with NADPH, CYP152A2 was able to convert 10 times more substrate, if provided with flavodoxin and flavodoxin reductase from *E. coli* and even 30–40 times more substrate with the CYP102A1-reductase from *Bacillus megaterium*. According to the clear preference for hydroxylation at α-position, CYP152A2 can be referred to as fatty acid α-hydroxylase.
© 2007 Elsevier Inc. All rights reserved.

Keywords: *Clostridium acetobutylicum*; P450 monooxygenase; Fatty acids; Peroxygenase; Activity reconstitution

The cytochrome P450s belong to a family of ubiquitous heme *b* containing monooxygenases that play pivotal roles in the detoxification of xenobiotics as well as in the secondary metabolism. P450 monooxygenases catalyze the introduction of one atom from molecular oxygen, with the other reduced to water. The basic P450 catalyzed reactions include hydroxylation of sp3-C atom, heteroatom oxygenation, epoxidation of double bond, and dealkylation (heteroatom release). Two electrons in form of hydrid ion required for P450 catalysis are delivered from NAD(P)H via flavoprotein and/or iron–sulfur redox partners. There are, however, P450 monooxygenases that do not need NAD(P)H and use peroxides for catalysis.

One would expect the presence of P450 monooxygenases exclusively in aerobes, since molecular oxygen is necessary for P450 functions. Nevertheless, the recent genome sequence of the anaerobic microorganism *Clostridium acetobutylicum* revealed a gene encoding for a P450 monooxygenase [1]. *C. acetobutylicum* is an endospore-forming microorganism, which is widely used for solvent production. Although bacteria belonging to the genus *Clostridium* are strictly anaerobic, they can tolerate microoxic conditions (up to 5% oxygen). As reported in the literature, molecular oxygen has a crucial effect on the growth of clostridia. However, the mechanism of growth inhibition and adaptive response to oxygen stress are not completely understood yet. Oxygen detoxification systems in clostridia include superoxide dismutases, superoxide reductases, peroxidases and rubrerythrin, which are responsible for reactive oxygen species (ROS) and intracellular peroxides scavenging [2–4]. Heme oxygenase in *C. tetani* and H$_2$O-forming NADH oxidase (noxA) in *C. aminovalericum* have

---

Abbreviations: P450, cytochrome P450 monooxygenase; IPTG, isopropyl-β-D-thiogalacto-pyranoside.
* Corresponding author. Fax: +49 711 685 631 96.
E-mail address: Vlada.Urlacher@itb.uni-stuttgart.de (V.B. Urlacher).

0006-291X/$ - see front matter © 2007 Elsevier Inc. All rights reserved.
doi:10.1016/j.bbrc.2007.07.155

also been reported as being involved in the establishment of an anoxic microenvironmental conditions [5,6]. Kawasaki et al. [6] reported that many proteins are directly and indirectly involved in oxygen response in *C. acetobutylicum* and suggested that not only oxygen metabolism but also active oxygen and lipid peroxide scavenging enzymes are important for the microoxic growth of this bacterium.

Sequence analysis revealed a 57% identity of the P450 enzyme from *C. acetobutylicum* with CYP152A1 (P450$_{Bs\beta}$) from *Bacillus subtilis* and therefore is considered as its ortholog. P450$_{Bs\beta}$ is a hydrogen peroxide driven fatty acid hydroxylase with mechanism different from most P450 monooxygenases [7,8]. A P450 monooxygenase that consumes molecular oxygen or hydrogen peroxide might also be involved in the oxygen metabolism in *C. acetobutylicum* and therefore is of potential interest from several points of view. Characterization of this enzyme would help to understand better its natural function in *C. acetobutylicum* and probably can give a rise for potential utility as a biocatalyst. Herein, we describe cloning and expression of CYP152A2 (in the following referred to as P450$_{CLA}$) from *C. acetobutylicum* in *Escherichia coli* and its biochemical characterization in terms of stability and substrate spectra.

Materials and methods

Chemicals and enzymes. All chemicals were purchased from Fluka (Buchs, Switzerland). Restriction endonucleases, T4 DNA ligase, Pfu DNA polymerase, IPTG, and SDS–PAGE protein ladder were obtained from Fermentas (St. Leon-Rot, Germany). NADPH was from Jülich Fine Chemicals (Jülich, Germany).

Cloning of P450$_{CLA}$ and P450$_{Bs\beta}$ encoding genes. All DNA manipulations and microbiological experiments were carried out by standard methods [9]. The gene encoding CYP152A2 in *C. acetobutylicum* ATCC 824 (GenBank Accession No. CAC3330) was amplified by PCR and cloned into pUC18 vector using primers 5′-GGGCTGCAGTTGGAA AAATAAAATATTTAATATATAAGAAAGGAGG-3′ and 5′-GGGG TCGACTATGATTAATGTAATGAATGTAAATTAATTCC-3′. Using this plasmid as template the P450 gene was further amplified by PCR using the oligonucleotide primers 5′-GCTAGCTAGCATGTTACTAAAAGA AAATAC-3′ and 5′-CCGCTCGAGTTAAAGCTTTAGATTAATATT ATC-3′. Genomic DNA of *B. subtilis* subsp. subtilis str. 168 was isolated using the standard phenol/chloroform precipitation protocol [9] and the *ybd*T-gene encoding CYP152A1 (GenBank Accession No. BSU02100) was amplified by PCR using primers 5′-CGGGATCCATGAATGAGCAGA TTCCACATG-3′ and 5′-CCGCTCGAGTTAACTTTTTCGTCTGATT CC-3′. Thermocycle program was as follows: 25 repetitions of 95 °C for 1 min followed by annealing at 53 °C and extension at 72 °C for 3 min. The amplified genes were purified and fully sequenced by automated DNA sequencing (GATC-Biotech, Germany). The PCR products were digested by endonucleases using appropriate restriction sites introduced during PCR (NheI and XhoI for P450$_{CLA}$ or BamHI and XhoI for P450$_{Bs\beta}$, respectively) and then ligated into the pET28a(+) expression vector (Merck Biosciences, Novagen, Darmstadt, Germany).

Overexpression and purification of P450 monooxygenases. The P450 genes were expressed in *E. coli* BL21 (DE3) cells harbouring a pET28a(+) plasmid containing the gene under control of the T7 phage promoter. Therefore, 200 μl of competent *E. coli* cells were transformed with the desired plasmid and the transformation mixture was used directly to inoculate 400 ml Luria-Bertani broth containing 30 μg/ml kanamycin. Cultures were grown overnight at 37 °C. Recombinant protein expression was then induced by addition of 1 mM IPTG (from 1 M stock in water), the incubation temperature was lowered to 30 °C and the culture was grown for 4 h. The cell pellet harvested from the culture by centrifugation was suspended in 6 ml of 50 mM Tris–HCl buffer, pH 7.5 and the cells were lysed by sonication on ice. Cell debris were removed by centrifugation at 20.000g for 20 min at 4 °C and the supernatant was recovered. Concentrations of the correctly folded P450 enzymes were estimated through CO differential spectra as described by Omura and Sato [10,11] using $\varepsilon_{450-490} = 91$ mM^{-1} cm^{-1}. UV/vis spectra were recorded at 24 °C on an Ultraspec 3100pro spectrophotometer (Amersham Biosciences, UK).

For purification, the crude protein extract was diluted in purification buffer (50 mM Tris–HCl, pH 7.5, containing 500 mM NaCl and 20 mM imidazol) and loaded on a Ni-NTA Superflow column (50 × 10 mm, Qiagen, Hilden, Germany) equilibrated with two column volumes purification buffer. Unspecifically bound proteins were removed by a washing step with two column volumes of purification buffer, containing 50 mM imidazol. The bound P450 was eluted with purification buffer, containing 200 mM imidazol and dialyzed in 2 l 50 mM Tris–HCl, pH 7.5, in order to remove imidazol.

Cloning and overexpression of CYP102A1-reductase. CYP102A1-reductase was expressed from the pET28a(+) plasmid carrying the part of CYP102A1 gene coding for a diflavin reductase. The gene fragment was amplified by PCR using the following primers: 5′-GCGGATCCATG AAAAGGCAGAAAACGC-3′ and 5′-CGGAATTCTACCCAGCCC ACACGTCTTTTGCG-3′ and cloned into pET28a(+) vector after digestion with BamHI and EcoRI endonucleases. The heterologous expression of the diflavin reductase was done as described for CYP102A1 holoenzyme previously [12].

Overexpression and purification of flavodoxin (Fld) and NADPH-flavodoxin reductase (Fpr). Fld and Fpr from *E. coli* were homologously expressed *E. coli* BL21 (DE3) from the pET11a plasmid containing a corresponding gene. Expression and purification were carried out as described elsewhere [13].

Spin-state shift and substrate binding constant determinations. Spin-state shifts upon substrate binding were assayed at 24 °C under aerobic conditions through substrate titrations by adding small (<5 μl) aliquots of a 5 mM stock of substrate or ligand dissolved in DMSO to a 1.5 μM P450 enzyme solution and recording spectral changes between 350 and 450 nm. An equal amount of DMSO was added into the reference sample for each measurement. The binding constants (K_D) were calculated by fitting the peak-to-through difference in absorbance to a Lineweaver–Burk type diagram using the Origin 7 program (Origin Labs).

Substrate conversion by P450 and product identification. Three different reconstituted enzyme systems were developed to measure substrate conversion by P450$_{CLA}$ and P450$_{Bs\beta}$. The first system utilizes hydrogen peroxide, while the second and third ones are based on the use of oxygen and NADPH. Activity measurements were carried out in two different setups: one set of experiments (setup A) was performed in 50 mM Tris–HCl buffer, pH 7.5, containing 0.15 μM P450, 60 μM substrate and 200 μM H$_2$O$_2$ (in the H$_2$O$_2$-based system) or alternatively 300 μM NADPH with 0.3 μM of each Fld and Fpr (second system) or 0.3 μM diflavin reductase of CYP102A1 (third system) in a total volume of 200 μl.

The other set of experiments (setup B) was carried out with higher enzyme concentrations, containing 2.5 μM P450 enzyme in a total volume in 50 mM Tris–HCl buffer, pH 7.5, and total volume of 500 μl. The H$_2$O$_2$-based system contained 25 μM substrate and 50 μM H$_2$O$_2$, while the NADPH-based systems were set up with 250 μM substrate and 500 μM NADPH and either 5 μM of each Fld and Fpr (second system) or 5 μM diflavin reductase of CYP102A1 (third system). As a negative control, P450-free cell lysate from *E. coli* harbouring an empty pET28a(+) vector was used.

Substrate conversion was carried out at 37 °C for either 1, 2, or 4 min (setup A) or for 60 min (setup B). Conversion was stopped by adding 20 μl 37% HCl and the reaction mixture was extracted twice with 500 μl diethyl ether. Stearic acid in final concentration 25 or 250 μM was added as internal standard. The combined organic phases were evaporated and the residue was dissolved in 35 μl *N,O*-bis(trimethylsilyl)trifluoroacetamide containing 1% trimethylchlorsilane. One microliter of this solution was analyzed by gas–liquid chromatography–mass spectroscopy (GC–MS) on a GC-MS QP2010 instrument (Shimadzu, Kyoto, Japan) equipped with a flame-ionization detector (FID) using a FS-Supreme-5 column (0.25 mm × 30 m, Chromatographie Service GmbH, Germany) and helium as carrier gas. The column temperature was controlled at 120 °C for 2 min. The temperature was then raised to 235 °C at the rate of 10 °C/min, and then to 300 °C at 30 °C/min.

Results and discussion

Expression of P450 enzymes

The P450$_{CLA}$-gene from *C. acetobutylicum* and the *ybd*T-gene of *B. subtilis* were amplified by PCR and cloned into standard expression vector pET28a(+). The determined nucleotide sequences of the cloned genes were in complete agreement with those in the genome database. P450$_{CLA}$ (CYP152A2) and P450$_{Bsβ}$ (CYP152A1) were expressed in soluble form in *E. coli* in reasonable yields (>25 mg/l). Both enzymes were isolated and purified on a Ni–NTA column using advantage of a His-Tag (Fig. 1 for P450$_{CLA}$). The two P450 enzymes showed the characteristic 448 nm absorption for the Fe^{II}(CO) complex with no evidence of the inactive P420 form. Spectroscopic characterization of P450$_{CLA}$ revealed spectra typical for a P450 monooxygenase (Fig. 2).

Substrate binding

Classical Type I substrate binding spectrum of P450 monooxygenases is characterized by a shift in the Soret band from 418 nm in the substrate free form to 390 nm for the high spin form when a substrate is bound in close proximity to the heme to displace the water bound to the iron. Therefore, this method was used for determination

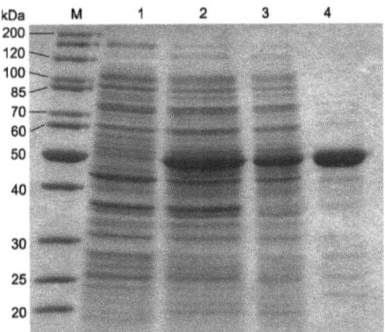

Fig. 1. SDS–polyacrylamide gel electrophoresis at various steps of P450$_{CLA}$ expression and purification. Lane 1, total cell protein before induction; lane 2, total cell protein 4 h after P450$_{CLA}$ expression was induced with 1 mM IPTG; lane 3, crude extract after sonication and centrifugation; lane 4, P450$_{CLA}$ with an estimated size of 48 kDa purified by Ni–NTA chromatography; M, PageRuler^{TM} Unstained Protein Ladder.

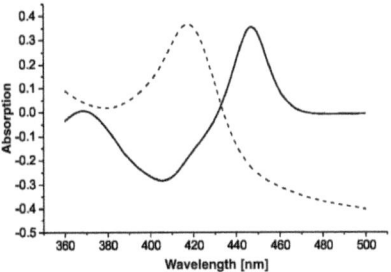

Fig. 2. Spectra of P450$_{CLA}$ in its different redox states: oxidized (dotted line); CO differential spectra (solid line).

of substrate spectra of P450$_{CLA}$. The screening included unsaturated and saturated fatty acids with chain lengths ranging from C_8 to C_{18}, several fatty acid esters, linear alkanes (C_8–C_{16}), and heteroaromatic and polycyclic aromatic compounds (Fig. 3).

No spin-state shift was induced in both P450 enzymes when chosen heteroaromatic and polycyclic aromatic compounds or alkanes were added to the enzymes, although P450$_{Bsβ}$ has recently been reported to accept anthracene, 9-methyl-anthracene and azulene [14].

P450$_{CLA}$ showed Type I shift with a variety of fatty acids (data not shown). The lowest K_D values were calculated for myristic, pentadecanoic and hexadecanoic acid. Fatty acids with shorter or longer chain length tend to weaken the binding, resulting in increasing K_D values (Table 1). In general, P450$_{CLA}$ was shown to bind a broad range of saturated and unsaturated fatty acids similar to P450$_{Bsβ}$. However, spin-state shifts for P450$_{CLA}$ could in general only be observed at higher substrate concentrations meaning that higher K_D values were calculated for this enzyme compared to P450$_{Bsβ}$ (Table 1).

We were surprised to see that spin-state shift could also be observed when the fatty acid was substituted with a corresponding methyl ester. For P450$_{Bsβ}$ it was previously reported that Arg242 (which is conserved in the P450$_{CLA}$ sequence as well) is a crucial residue for substrate binding,

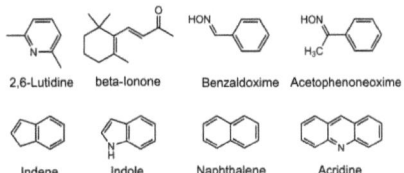

Fig. 3. Heteroaromatic and polycyclic aromatic substrates assayed for P450$_{CLA}$ and P450$_{Bsβ}$ in this study.

Table 1
K_D values for P450$_{CLA}$ and P450$_{Bsβ}$ determined from substrate induced Type I spin-state shifts

Substrate	K_D value for P450$_{CLA}$ (µM)	K_D value for P450$_{Bsβ}$ (µM)
Caprylic acid	>1000	422 ± 27
Capric acid	>1000	414 ± 25
Lauric acid	221 ± 22	88 ± 15
Tridecanoic acid	164 ± 72	15 ± 2
Myristic acid	36 ± 5	9 ± 0.5
Pentadecanoic acid	25 ± 5	8.5 ± 0.5
Palmitic acid	30 ± 6	10 ± 0.5
Heptadecanoic acid	98 ± 21	8.5 ± 1.5
Stearic acid	98 ± 31	40 ± 17
Oleic acid	33 ± 3	22 ± 3
Linolenic acid	70 ± 14	34 ± 4
Arachidonic acid	n.d.	50 ± 5
Stearic acid methylester	690 ± 100	1220 ± 440
Palmitic acid methylester	105 ± 10	72 ± 18
Myristic acid methylester	74 ± 13	30 ± 3
Lauric acid ethylester	(−)	(−)
Lauric acid butylester	(−)	(−)
n-Octane	(−)	(−)
Decane	(−)	(−)
Dodecane	(−)	(−)
Hexadecane	(−)	(−)
Cyclohexene	(−)	(−)
Aromatic compounds (see Fig. 3)	(−)	(−)

n.d., not determined; (−), no spin-state shift observed.

interacting with the negatively charged carboxy group of the fatty acid via hydrophobic interaction and thus determining the regioselectivity of this enzyme. Mutation of this residue by site-directed mutagenesis results in a great inhibition of hydroxylation of myristic acid [15,16]. However, a weakening of this interaction by substitution of the carboxy group with a methylester group still allows the substrate to bind, although with lower affinity compared to the corresponding fatty acid (Table 1). Interestingly, when the carboxy group was substituted with an ethylester or butylester, a spin-state shift could not be observed.

Activity reconstitution and identification of oxidation products by GC–MS

Conversion of lauric, myristic or palmitic acid by P450$_{CLA}$ was investigated with 200 µM H$_2$O$_2$. GC–MS analysis demonstrated that P450$_{CLA}$ was able to catalyze the hydroxylation of fatty acids to produce the α- and β-hydroxylated derivatives with quite high initial activities (Table 2). After 2 min of reaction conversion of 60 µM myristic acid reached 40%, lauric acid—55%, and palmitic acid—60%. However, product formation was observed only in the first minutes of the reaction. Presumably, the high concentration of H$_2$O$_2$ led to a rapid inactivation of enzyme resulting in a complete loss of activity (data not shown).

To increase stability of the system, we increased enzyme concentration (2.5 µM) and decreased H$_2$O$_2$ concentration (50 µM). This resulted, however, in worse productivity: after 1 h only 8% of 25 µM lauric acid, 34% of myristic acid, and 10% of palmitic acid were converted (Table 3). Thus, a high excess of hydrogen peroxide is necessary for high enzyme activity.

To study whether P450$_{CLA}$ accepts oxygen we used two NADPH-based reaction systems to reconstitute the activity of P450$_{CLA}$ and replace the electron donor H$_2$O$_2$. There is no evidence for P450$_{CLA}$ or P450$_{Bsβ}$ belonging to a certain type of P450 in terms of electron partner. Moreover, for P450$_{Bsβ}$ it is reported that ferredoxin with ferredoxin reductase and cytochrome P450 reductase systems do not appear to function with this enzyme [7]. Nevertheless, we

Table 2
Initial activities[a] of P450$_{CLA}$ and P450$_{Bsβ}$, measured within 2 min using setup A

Electron donor and substrate	H$_2$O$_2$			Fld and Fpr from *E. coli*	CYP102A1-reductase		
	Lauric acid	Myristic acid	Palmitic acid	Myristic acid	Lauric acid	Myristic acid	Palmitic acid
P450$_{CLA}$	108.2	194.8	115.4	(−)	(−)	5.9	1.0
P450$_{Bsβ}$	38.3	72.2	15.6	(−)	9.9	24.1	14.5

(−), no activity observed.
[a] The activity was calculated in nmol substrate per min per nmol P450.

Table 3
Substrate conversion by P450$_{CLA}$ and activity[a] measured within 60 min using setup B

Electron acceptor and substrate	H$_2$O$_2$ ($c_{Substrate}$ = 25 µM)			Fld and Fpr from *E.coli* ($c_{Substrate}$ = 250 µM)			CYP102A1-reductase ($c_{Substrate}$ = 250 µM)		
	Lauric acid	Myristic acid	Palmitic acid	Lauric acid	Myristic acid	Palmitic acid	Lauric acid	Myristic acid	Palmitic acid
Substrate [%][b]	92.1	66.3	90.1	89.6	83.0	88.9	58.2	39.1	69.7
α-Product [%][b]	7.9	22.1	4.7	7.2	11.3	9.8	34.5	51.0	26.4
β-Product [%][b]	0.0	11.6	5.1	3.3	5.7	1.3	7.3	9.9	3.9
Activity[a]	0.8	3.4	1.0	10.4	16.7	11.1	41.8	66.3	30.1

The data are averages of at least three experiments, with standard deviations within ~10% of the mean.
[a] Activity was determined as nmol substrate per h per nmol protein.
[b] %-values correspond to the percental amount of the substance determined by peak integration via GC-MS.

investigated substrate conversion by P450$_{CLA}$ with reconstituted enzyme systems containing either flavodoxin (Fld) and flavodoxin reductase (Fpr) from *E. coli*, which have been described to be structurally similar to the functional domains (FMN binding and NADPH/FAD binding domains, respectively) of NADPH–cytochrome P450 reductases [13] or the diflavin reductase domain of CYP102A1 (P450 BM3) from *Bacillus megaterium*.

Table 4
Substrate conversion by P450$_{B\mu\beta}$ and activity[a] measured within 60 min using setup B

Electron acceptor and substrate	H$_2$O$_2$ ($c_{Substrate}$ = 25 μM)			Fld and Fpr from *E.coli* ($c_{Substrate}$ = 250 μM)			CYP102A1-reductase ($c_{Substrate}$ = 250 μM)		
	Lauric acid	Myristic acid	Palmitic acid	Lauric acid	Myristic acid	Palmitic acid	Lauric acid	Myristic acid	Palmitic acid
Substrate [%][b]	13.6	1.8	36.1	7.2	5.7	24.8	2.6	2.9	21.5
α-Product [%][b]	12.4	7.0	6.6	31.8	13.8	15.3	34.2	17.4	16.5
β-Product [%][b]	74.0	91.2	57.2	61.0	80.5	59.9	63.2	79.7	62.0
Activity[a]	8.6	9.8	6.4	92.8	94.3	75.2	97.4	97.1	78.5

The data are averages of at least three experiments, with standard deviations within ~10% of the mean.
[a] Activity was determined as nmol substrate per h per nmol protein.
[b] %-values correspond to the percental amount of the substance determined by peak integration via GC–MS.

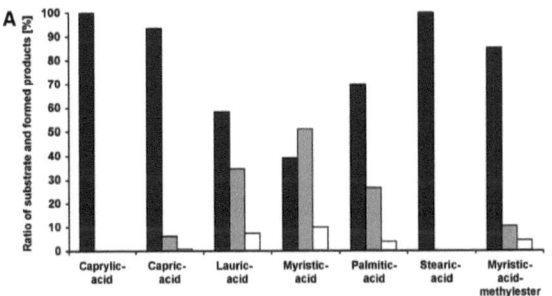

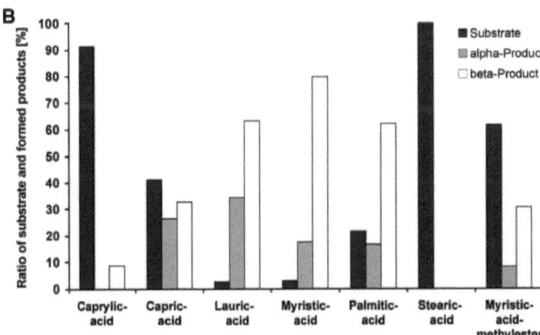

Fig. 4. Hydroxylation of fatty acids by P450$_{CLA}$ (A) or P450$_{B\mu\beta}$ (B) using the CYP102A1-reductase reconstituted enzyme system and setup B. After incubation at 37 °C for 1 h, the reaction mixture was analyzed by GC–MS and the ratio of substrate and formed products was determined by peak integration. The data are averages of two experiments, with standard deviations within ~5% of the mean.

The initial activities of P450$_{CLA}$ in both reconstitution systems were either much lower than those measured with H$_2$O$_2$ or not measurable at all (Table 2). Nevertheless, since stability of the enzyme was not affected by H$_2$O$_2$ in these systems, product formation after one hour was significantly higher than with H$_2$O$_2$ (Table 3). P450$_{CLA}$ was able to convert 10 times more substrate, if provided with Fld and Fpr and even 30–40 times more substrate, if CYP102A1-reductase was added to the reaction mixture. Remarkably, the same tendency was observed for P450$_{BsB}$ (Table 4).

For P450$_{CLA}$ the ratio of α- to β-derivatives was ranging from 2:1 to 5:1 (depending on the substrate and conditions chosen). Thus according to the clear preference for hydroxylation at α-position the P450$_{CLA}$ enzyme can be referred to as fatty acid α-hydroxylase.

Using the reconstituted CYP102A1-reductase system we determined the minimal and maximal chain length that is necessary for substrate recognition by both P450 enzymes. Further, we also investigated, if myristic acid methylester can be accepted as substrate. For P450$_{CLA}$ no conversion was detected for caprylic acid (C$_8$) and stearic acid (C$_{18}$) and thus even chain fatty acids with chain length between C$_{10}$ and C$_{16}$ are the substrates for P450$_{CLA}$ (Fig. 4A). P450$_{BsB}$ was able to convert small amounts of caprylic acid (Fig. 4B), but was not able to convert stearic acid as well. Both enzymes showed conversion of myristic acid methylester to its α- and β-derivatives. The ratio of α- and β-product was the same as for myristic acid, but after one hour of conversion, 55% or 40% less product was detected for P450$_{CLA}$ or P450$_{BsB}$, respectively.

Conclusions

According to its protein sequence, P450$_{CLA}$ is an ortholog of the peroxygenase P450$_{BsB}$. Indeed, P450$_{CLA}$ demonstrated high activity against fatty acids in the presence of hydrogen peroxide. However, this activity could only be observed within the first few minutes of the reaction. This can be explained by the natural function of P450$_{CLA}$ as an enzyme being involved in hydrogen peroxide scavenging in *C. acetobutylicum*. On the other hand, its role as a member of the oxygen detoxification system can be more complicated, as the enzyme accepts molecular oxygen as well. In this case it would need redox partner(s), providing electrons for catalysis. A flavodoxin is a likely candidate for a physiological electron donor and its identification would provide important information for understanding the role of the P450 enzyme in the anaerobic microorganism.

Acknowledgments

M.G. and V.B.U. thank Kyoko Momoi for the help in preparation of Fld and Fpr. P.D. and S.S. acknowledge funding by the BMBF projects GenoMikPlus (Competence Network Göttingen) and SysMO COSMIC (PtJ-BIO/SysMO/P-D-01-06-13), www.sysmo.net.

References

[1] J. Nölling, G. Breton, M.V. Omelchenko, K.S. Makarova, Q. Zeng, R. Gibson, H.M. Lee, J. Dubois, D. Qiu, J. Hitti, Y.I. Wolf, R.L. Tatusov, F. Sabathe, L. Doucette-Stamm, P. Soucaille, M.J. Daly, G.N. Bennett, E.V. Koonin, D.R. Smith, Genome sequence and comparative analysis of the solvent-producing bacterium *Clostridium acetobutylicum*, J. Bacteriol. 183 (2001) 4823–4838.

[2] D. Jean, V. Briolat, G. Reysset, Oxidative stress response in *Clostridium perfringens*, Microbiology 150 (2004) 1649–1659.

[3] P. Dürre, Clostridia, Encyclopedia Life Sci. (2007), doi:10.1002/9780470015902, a9780470020370.

[4] F. Hillmann, R.J. Fischer, H. Bahl, The rubrerythrin-like protein Hsp21 of *Clostridium acetobutylicum* is a general stress protein, Arch. Microbiol. 185 (2006) 270–276.

[5] H. Brüggemann, R. Bauer, S. Raffestin, G. Gottschalk, Characterization of a heme oxygenase of *Clostridium tetani* and its possible role in oxygen tolerance, Arch. Microbiol. 182 (2004) 259–263.

[6] S. Kawasaki, Y. Watamura, M. Ono, T. Watanabe, K. Takeda, Y. Niimura, Adaptive responses to oxygen stress in obligatory anaerobes *Clostridium acetobutylicum* and *Clostridium aminovalericum*, Appl. Environ. Microbiol. 71 (2005) 8442–8450.

[7] D.S. Lee, A. Yamada, H. Sugimoto, I. Matsunaga, H. Ogura, K. Ichihara, S. Adachi, S.Y. Park, Y. Shiro, Substrate recognition and molecular mechanism of fatty acid hydroxylation by cytochrome P450 from *Bacillus subtilis*. Crystallographic, spectroscopic, and mutational studies, J. Biol. Chem. 278 (2003) 9761–9767.

[8] I. Matsunaga, A. Ueda, N. Fujiwara, T. Sumimoto, K. Ichihara, Characterization of the ybdT gene product of *Bacillus subtilis*: novel fatty acid beta-hydroxylating cytochrome P450, Lipids 34 (1999) 841–846.

[9] J. Sambrook, D.W. Russel, Molecular Cloning: A Laboratory Manual, third ed., Cold Spring Harbor Laboratory Press, New York, 2001.

[10] T. Omura, R. Sato, The carbon monoxide-binding pigment of liver microsomes. Ii. Solubilization, purification, and properties, J. Biol. Chem. 239 (1964) 2379–2385.

[11] T. Omura, R. Sato, The carbon monoxide-binding pigment of liver microsomes. I. Evidence for its hemoprotein nature, J. Biol. Chem. 239 (1964) 2370–2378.

[12] S. Maurer, A. Urlacher, H. Schulze, R.D. Schmid, Immobilisation of P450 BM-3 and an NADP+ cofactor recycling system: towards a technical application of heme-containing monooxygenases in fine chemical synthesis, Adv. Synth. Catal. 345 (2003) 802–810.

[13] C.M. Jenkins, M.R. Waterman, NADPH-flavodoxin reductase and flavodoxin from *Escherichia coli*: characteristics as a soluble microsomal P450 reductase, Biochemistry 37 (1998) 6106–6113.

[14] E. Torres, H. Hayen, C.M. Niemeyer, Evaluation of cytochrome P450(BSbeta) reactivity against polycyclic aromatic hydrocarbons and drugs, Biochem. Biophys. Res. Commun. 355 (2007) 286–293.

[15] I. Matsunaga, T. Sumimoto, M. Ayata, H. Ogura, Functional modulation of a peroxygenase cytochrome P450: novel insight into the mechanisms of peroxygenase and peroxidase enzymes, FEBS Lett. 528 (2002) 90–94.

[16] I. Matsunaga, A. Ueda, T. Sumimoto, K. Ichihara, M. Ayata, H. Ogura, Site-directed mutagenesis of the putative distal helix of peroxygenase cytochrome P450, Arch. Biochem. Biophys. 394 (2001) 45–53.

2.2.2. Manuscript: Expression, purification and characterization of two *Clostridium acetobutylicum* flavodoxins: Potential electron transfer partners for CYP152A2

Material from this chapter appears in:
Sumire Honda Malca[a], Marco Girhard[b], Stefanie Schuster[c], Peter Dürre[c] and Vlada B. Urlacher[b], 2010, Expression, purification and characterization of two *Clostridium acetobutylicum* flavodoxins: Potential electron transfer partners for CYP152A2, *Biochimica et Biophysica Acta*, doi:10.1016/j.bbapap.2010.06.013.

[a] Institute of Technical Biochemistry, Universität Stuttgart, 70569 Stuttgart, Germany
[b] Institute of Biochemistry, Heinrich-Heine-University Düsseldorf, 40225 Düsseldorf, Germany
[c] Institute of Microbiology and Biotechnology, University of Ulm, 89069 Ulm, Germany
* Corresponding author

Material is reprinted by permission of Elsevier; the original manuscript is available online at: http://www.sciencedirect.com/science/journal/15709639

ARTICLE IN PRESS

Biochimica et Biophysica Acta xxx (2010) xxx-xxx

Contents lists available at ScienceDirect

Biochimica et Biophysica Acta

journal homepage: www.elsevier.com/locate/bbapap

Expression, purification and characterization of two *Clostridium acetobutylicum* flavodoxins: Potential electron transfer partners for CYP152A2

Sumire Honda Malca [a], Marco Girhard [b], Stefanie Schuster [c], Peter Dürre [c], Vlada B. Urlacher [b],*

[a] *Institute of Technical Biochemistry, Universität Stuttgart, 70569 Stuttgart, Germany*
[b] *Institute of Biochemistry, Heinrich-Heine-University Düsseldorf, 40225 Düsseldorf, Germany*
[c] *Institute of Microbiology and Biotechnology, University of Ulm, 89069 Ulm, Germany*

ARTICLE INFO

Article history:
Received 19 March 2010
Received in revised form 10 June 2010
Accepted 17 June 2010
Available online xxxx

Keywords:
Clostridium acetobutylicum
Cytochrome P450 monooxygenase
CYP152A2
Flavodoxin
Activity reconstitution

ABSTRACT

Two flavodoxin genes from *Clostridium acetobutylicum*, CacFld1 (CAC0587) and CacFld2 (CAC3417), were expressed in *Escherichia coli* and investigated for their ability to support activity of CYP152A2, a fatty acid hydroxylase from *C. acetobutylicum*. *E. coli* flavodoxin reductase (FdR) was used as a redox partner, since flavodoxin reductase CacFdR (CAC0196) from *C. acetobutylicum* could not be purified in a functional form. CacFld1 was shown to accept electrons from FdR and transfer them to CYP152A2. Since H_2O_2 was generated by uncoupling at different stages of the reconstituted electron transfer chain, catalase was used as H_2O_2 scavenger in order to exclude peroxygenation by CYP152A2. The reconstituted P450 system with CacFld1 and FdR oxidized myristic acid with a K_M of 137 μM and a k_{cat} of 36 min^{-1}. Furthermore, the hydroxylase activity of CYP152A2 towards myristic acid with CacFld1 was 17-fold higher than without CacFld1. Along with CYP152A2 and a physiological flavodoxin reductase, CacFld1 is therefore likely to be involved in oxygen detoxification in *C. acetobutylicum*. Flavodoxin CacFld2 did not accept electrons from NADPH-reduced FdR, though it cannot be excluded as a candidate redox partner for CYP152A2 in the presence of an appropriate physiological reductase.

© 2010 Elsevier B.V. All rights reserved.

1. Introduction

Cytochromes P450 (P450s or CYPs) are heme-iron containing oxidoreductases found in all domains of life [1,2]. P450s catalyze monooxygenase reactions, which introduce one atom of molecular oxygen into generally non-polar, aromatic or aliphatic molecules, thereby leading to hydroxylation, aromatization, epoxidation or cleavage of carbon—carbon bonds. This process requires the consecutive delivery of two reducing equivalents, derived from the nicotinamide cofactors NADH or NADPH and transferred to the heme iron via redox partners [3]. Thus most P450 systems are composed of the P450 itself and one or two additional proteins or protein domains constituting an electron transfer chain. Depending on the redox partners, traditionally two main classes of P450s have been defined [4]: Class I P450s are found in mitochondria of eukaryotes and in bacteria. They use a small [2Fe–2S] protein (ferredoxin) and a FAD-containing reductase (ferredoxin reductase) for the transfer of electrons from NAD(P)H to the P450. Class II P450s

Abbreviations: CacFld1, *C. acetobutylicum* flavodoxin CAC0587; CacFld2, *C. acetobutylicum* flavodoxin CAC3417; CacFdR, *C. acetobutylicum* flavodoxin reductase CAC0196; FdR, *E. coli* flavodoxin (ferredoxin) NADP(H)-dependent oxidoreductase
* Corresponding author. Universitätsstraße 1, 40225 Düsseldorf, Germany. Tel.: +49 211 81 13889; fax: +49 211 81 13117.
E-mail address: Vlada.Urlacher@uni-duesseldorf.de (V.B. Urlacher).

rely on FAD- and FMN-containing cytochrome P450 reductases (CPR) for the transfer of electrons from NADPH. Mammalian microsomal and plant P450s, as well as some fungal P450s belong to this class.

In recent years numerous genome sequencing projects have revealed many other types of electron transfer proteins, which belong neither to class I nor to class II. Consequently the "old" P450 classification system has been reorganized [3]. According to the redox partner diversity, 10 different P450 classes can be distinguished, however, some of these classes are rare and represented by only one or two examples. For instance, the novel class III includes so far (strictly speaking) only CYP176A1 (P450cin) from *Citrobacter braakii* [5]. This three-component system consists of an FAD-containing ferredoxin reductase, an FMN-containing flavodoxin and the P450. Concerning the redox centers, the P450cin system resembles the eukaryotic microsomal system. But unlike a diflavin CPR, in this novel system the two redox centers FAD and FMN belong to separate proteins. A flavodoxin from *C. braakii* was designated as cindoxin. Therefore, in this new bacterial system the electrons are delivered via the redox centers FAD and FMN and not via FAD and an iron–sulfur-cluster as in the class I systems.

Flavodoxins are small electron transfer proteins (typically 14–23 kDa) that contain one molecule of non-covalently but tightly bound FMN as the redox active component. They were first discovered in cyanobacteria [6] and clostridia [7] growing in low-iron conditions, where they replaced the iron-containing ferredoxin in reactions leading to NADP$^+$ and N$_2$-reduction. Although these remain perhaps their best-known physiological roles, flavodoxins are involved in a variety of other important reactions and in some organisms they are essential, constitutive proteins [8]. For example, there is evidence for the involvement of a flavodoxin as the electron donor in the reaction of *Escherichia coli* methionine synthase [9].

It has previously been reported that flavodoxin reductase and flavodoxin from *E. coli* can substitute for the endogenous redox partners of heterologously expressed cytochrome P450s [10,11]. Later it has been demonstrated that at least one of two flavodoxins (YkuN and YkuP) from *Bacillus subtilis* can serve as redox partner for CYP107H (P450BioI) [12] and CYP109B1 [13], both originate from the same strain. Nevertheless, it could not be ruled out that these P450s belong to the bacterial class I or III P450s, since there are other experimental data demonstrating that they also interact with ferredoxins [13–15].

Recently we have characterized the cytochrome P450 monooxygenase CYP152A2 (P450CLA) from *Clostridium acetobutylicum* [16]. CYP152A2 has been classified to so-called peroxygenases that do not involve the "classical" P450 reaction cycle, but the peroxide-shunt, which is the reversal of the uncoupling reaction that leads to the collapse of the ferric hydroperoxy species. Other enzymes belonging to the CYP152-family are the fatty acid hydroxylating peroxygenases CYP152B1 (SPα) from *Sphingomonas paucimobilis* [17] and CYP152A1 (P450Bsβ) from *B. subtilis* [18].

Clostridia are classified as anaerobic Gram-positive bacteria. Certain species such as *Clostridium tetani*, *Clostridium perfringens*, and *Clostridium botulinum* are toxin-producing microorganisms responsible for life-threatening diseases, whereas a large number of non-pathogenic clostridia such as *C. acetobutylicum*, *Clostridium beijerenckii*, and *Clostridium kluyveri* have been used in solvent fermentation, biodegradation, and investigation of microbial energy conservation [19–21]. Despite their classification as anaerobes, many species are able to grow in a microoxic environment. This suggests that clostridia possess some adaptive systems to metabolize oxygen, as well as to scavenge reactive oxygen species (ROS) derived from oxygen reduction [22]. Although the mechanisms involved in oxygen tolerance in clostridia have not been elucidated yet, proteins such as superoxide dismutase, superoxide reductase, catalase, glutathione peroxidase, thioredoxin/thioredoxin reductase, NAD(P)H-dependent alkyl hydroperoxide reductase, NAD(P)H oxidase, rubrerythrin, heme-oxygenase, and peroxide repressor have been associated with ROS detoxification [22–28].

CYP152A2 was demonstrated to have high fatty acid oxidation activity by consuming either hydrogen peroxide or molecular oxygen, which might reflect the natural function of this P450 as a ROS-scavenging enzyme in *C. acetobutylicum*. The activity of CYP152A2 was successfully reconstituted in two different systems with molecular oxygen and NADPH. Electron transfer partners were represented either by flavodoxin and flavodoxin reductase from *E. coli* or by the diflavin reductase domain of CYP102A1 from *Bacillus megaterium* [16]; however, physiological redox partners of CYP152A2 have not been identified yet.

Flavodoxins and flavodoxin oxidoreductases are likely candidates for natural electron donors of CYP152A2. At least 27 genes encoding flavodoxin-like proteins and oxidoreductases including flavin- or iron-sulfur-containing proteins from the genome of *C. acetobutylicum* ATCC 824 have been identified or predicted by genome sequencing and gene annotation [20]. Previous studies have focused on the recombinant expression of *C. acetobutylicum* flavodoxin CAC0587 [29] and ferredoxins CAC0303 and CAC3527 [30] as electron transfer partners of [Fe-Fe]-hydrogenase. However, none of them was considered as a potential redox partner of a P450 enzyme.

The aim of the present work was to identify flavoproteins that could serve as potential redox partners for CYP152A2. Two flavodoxin genes – CAC0587 (*Cac*Fld1) and CAC3417 (*Cac*Fld2) – and a flavodoxin reductase – CAC0196 (*Cac*FdR) – were cloned and expressed in *E. coli* and tested for their ability to reconstitute the activity of CYP152A2 *in vitro*. The characterization of these redox proteins provides important information for understanding the physiological role of CYP152A2 in *C. acetobutylicum*.

2. Materials and methods

2.1. Chemicals, enzymes and strains

Pfu DNA polymerase, restriction endonucleases, T4 DNA ligase, IPTG, DNA Ladder were obtained from Fermentas (St. Leon-Rot, Germany) and used as recommended by the manufacturer. Horse heart cytochrome *c* and bovine liver catalase were procured from Fluka. Glucose-6-phosphate dehydrogenase from *S. cerevisiae* was from Roche Diagnostics (Mannheim, Germany). NADPH tetrasodium salt was from Codexis (Jülich, Germany). The "Qiagen PCR cloning kit" including the vector pDrive was obtained from Qiagen Gmbh (Hilden, Germany). *E. coli* strain BL21(DE3) and vectors pET28a(+) and pET16b were obtained from Novagen (Madison, Wisconsin, USA). *C. acetobutylicum* ATCC 824 stemmed from the Ulm laboratory collection. All other chemicals, solvents and buffer components were purchased from Sigma-Aldrich (Schnelldorf, Germany).

2.2. Cloning, expression and purification of flavoproteins

Genes encoding flavodoxins *Cac*Fld1 (GeneBank CAC0587) and *Cac*Fld2 (GeneBank CAC3417), as well as the flavodoxin oxidoreductase *Cac*FdR (GeneBank CAC0196) were derived from genomic DNA of *C. acetobutylicum* ATCC 824 and cloned into plasmid pDrive as follows: Purified PCR products obtained by the "Qiagen PCR cloning kit" contain a 5′ A-overhang, which can be used for direct cloning into the linearized vector (containing a 3′ T-overhang). Ligation was performed using 100 ng purified PCR fragment. Further subcloning was possible by using the introduced restriction sites for *Pst*I and *Sal*I in designed primers. The following primers were used (restriction sites are written in lower case letters): 5′-GGGctgcagTTAAATATGTAGGAGGAG-3′ (cac0587*Pst*Ifw) and 5′-GGCgtcgacTTCCATCCTCTATCTCTC-3′ (cac0587*Sal*Irev) for CAC0587; 5′-GGGctgcagAGTTAAATAAGGAGTAAG-3′ (cac3417*Pst*Ifw) and 5′-GGCgtcgacCACTGCGCACAATGTATG-3′ (cac3417*Sal*Irev) for CAC3417; 5′-GGGctgcagATATAATAATAACATACC-3′ (cac0196*Pst*Ifw) and 5′-GGCgtcgacCACAGCACTACTAGCTTC-3′ (cac0196*Sal*Irev) for CAC0196.

The two flavodoxin genes were recloned into pET28a(+) for recombinant expression in *E. coli* as C-terminal His-tagged proteins. *Cac*Fld1 was amplified using oligonucleotides 5′-GGATccatggTGAAAA-TAAACATAATTTATTGC-3′ and 5′-GATCctcgagGCTATTTATAAGAGCCTT-3′, *Cac*Fld2 with oligonucleotides 5′-GGTccatggGCATGAAAAGTTTAA-TAG-3′ and 5′-GGCCgctcgagTTTTTCTAAATACTTTGTTA-3′. PCR was performed with 26 cycles at 95 °C for 1 min; 58 °C for 30 s, 72 °C for 1.5 min. Amplified products were digested by endonucleases *Nco*I and *Xho*I and ligated into previously linearized pET28a(+).

*Cac*FdR was cloned into pET16b using the primers 5′-GGTcccatggGCATAAAAGTTTAATAG-3′ and 5′-GATCggatccTTAGT-GATGATGATGATGGGCATTATTTCT-3′ (which introduced a C-terminal His$_6$-tag coding sequence; underlined).

The FdR-encoding gene *fpr* from *E. coli* K-12 strain JM109 was cloned utilizing the plasmid pET11a-*fpr* previously described [31] as template for PCR. Primers 5′-GTATccatgggcCATCATCATCATCAT-CATGCTGATTGGGTA-3′ (encoding an N-terminal His$_6$-tag sequence; underlined) and 5′-GATAggatccTTACCAGTAATGCTCCGCTGTCAT-3′ were used for cloning of FdR into pET16b. The two flavodoxin reductase genes were PCR-amplified according to the cycle program described above. Products were digested by *Nco*I and *Bam*HI and ligated into previously linearized pET16b. Sequences of all inserts in the resulting plasmids were verified by automated DNA-sequencing (GATC-Biotech, Konstanz, Germany).

Please cite this article as: S. Honda Malca, et al., Expression, purification and characterization of two *Clostridium acetobutylicum* flavodoxins: Potential electron transfer partners ..., Biochim. Biophys. Acta (2010), doi:10.1016/j.bbapap.2010.06.013

Protein expression was carried out following adapted protocols described in [29] and [31]. For expression of Flds and FdRs, cells were grown at 37 °C in LB with 30 μg/μl kanamycin or TB containing 1 mM $MgCl_2$, 5 mM NaCl, and 100 μg/μl ampicillin, until reaching an OD_{600} of 0.5 for induction with 20 μM (Flds) or 100 μM IPTG (FdRs), followed by incubation at 25 °C, 140 rpm. After 16 h, cells were collected and suspended in 50 mM Tris–HCl pH 7.5, 100 μM PMSF, 5% glycerol; incubated with lysozyme (0.5 mg/ml) at 4 °C for 30 min and disrupted by sonication (10 × 20 s, 1.5 min intervals). Protein purification was done using a Ni-NTA affinity column (Qiagen GmbH, Germany) preequilibrated with 50 mM Tris–HCl pH 7.5, 100 μM PMSF, 5% glycerol, and 300 mM NaCl. Flavoproteins were eluted with the same buffer containing 150 mM imidazole and dialyzed for 4 h at 4 °C against 50 mM Tris–HCl pH 7.5, 100 μM PMSF, 5% glycerol. Following dialysis, protein solutions were concentrated by ultrafiltration and stored in aliquots at −20 °C. The purity of Flds and FdRs was checked by SDS-PAGE on 15% and 12.5% polyacrylamide gels, respectively.

2.3. Characterization of flavoproteins

Spectral properties of the purified flavoproteins were analyzed between 300 and 600 nm on a UV/Vis Ultrospec 3000 (Pharmacia Biotech, Sweden). Flavodoxin and CacFdR concentrations were determined by measuring FMN or FAD via fluorescence spectroscopy according to a method described previously [32]. The flavin cofactor was released from the holoflavoprotein by incubating 100 μl of protein solution with 25 μl of 25% (v/v) trichloroacetic acid for 5 min (25 °C, dark conditions). Samples were centrifuged (12,000g for 10 min, 4 °C), the supernatant fractions were collected and neutralized with 75 μl of 2 M K_2HPO_4. The concentration of cofactor in the supernatant was determined from its fluorescence (excitation at 470 nm and emission at 510 nm) [33] and calculated by using a series of FMN and FAD standard solutions in 50 mM Tris–HCl pH 7.5, treated under the same conditions as the samples.

The concentration of purified FdR was determined by UV/Vis spectroscopy using $\varepsilon_{456} = 7.1$ mM^{-1} cm^{-1} [34].

2.4. Expression and purification of CYP152A2

CYP152A2 (GenBank CAC3330) derived from *C. acetobutylicum* ATCC 824 was cloned into pET28a(+) as described previously [16]. Expression, purification and quantification were performed as explained therein.

2.5. Reduction of flavodoxins by FdR

A solution containing 6.25 μM Fld (CacFld1 or CacFld2) and 1.25 μM FdR was prepared in 50 mM Tris–HCl pH 7.5. The solution was added to a 0.6-ml quartz cuvette, flushed with nitrogen, capped, and its UV-visible spectrum (300–700 nm) recorded. NADPH was then added to a final concentration of 12.5 μM and the spectrum monitored every 30 s for 5 min.

Cytochrome *c* reductase activity of FdR with and without CacFld1 was followed at 550 nm ($\varepsilon = 21.1$ mM^{-1} cm^{-1}) [35] in 50 mM Tris–HCl pH 7.5 containing 1 μM FdR, 5 μM CacFld1, 100 μM cytochrome *c*, and 500 μM NADPH.

2.6. Reduction of P450 heme iron by FdR and CacFld1

Reduction of the P450 heme iron was monitored in NADPH/FdR/P450 systems with and without CacFld1 utilizing the CO differential spectral method described by Omura and Sato [36,37]. Solutions containing 3 μM CYP152A2, 15 μM of each flavoprotein, 100 μM myristic acid (from a 10 mM stock solution in DMSO), and 1 mM EDTA in 50 mM Tris–HCl pH 7.5 were made anaerobic by flushing with N_2 for 30 min on ice [38]. NADPH was added from a stock solution previously gassed with N_2 to a final concentration of 500 μM. The solutions were divided into N_2-gassed quartz cuvettes and sealed with Teflon covers. Samples were saturated with carbon monoxide by bubbling with gas for 1 min. Spectral measurements were recorded between 400 and 500 nm over periods of 5 min. A second experiment included the use of an oxygen scavenger system employing glucose oxidase/catalase (GOD/CAT) (3000 U/ml catalase, 0.1 U/ml glucose oxidase, and 100 mM β-D-glucose) [39] simultaneously with N_2-flushing to maintain the anaerobic environment. Samples were subjected to the same conditions as described above and spectra were recorded every 5 min. To exclude H_2O_2 as the cause for P450 heme-iron reduction, 100 μM H_2O_2 was added to a solution containing CYP152A2 and myristic acid in the concentrations given above. The solution was purged with nitrogen, capped, and the spectrum was recorded every 5 min for 1 h.

2.7. CYP152A2 hydroxylase activity with FdR and CacFld1 as redox partners

Hydroxylation of myristic acid by CYP152A2 was carried out in 50 mM Tris–HCl pH 7.5 in a final volume of 300 μl, and in the presence of catalase (for H_2O_2 scavenging) and glucose-6-phosphate dehydrogenase (for NADPH regeneration). The concentrations used were 0.25 μM CYP152A2, 1.25 μM FdR, 6.25 μM CacFld1, 600 U/ml bovine liver catalase, 5 mM glucose-6-phosphate (G6P), 1 mM $MgCl_2$, 2.75 U/ml glucose-6-phosphate dehydrogenase (G6PDH). Myristic acid in a final concentration of 200 μM was added from a 10 mM stock solution in DMSO. The reaction was started by the addition of 300 μM NADPH, followed by incubation at 37 °C. Conversion was stopped after 1 h by addition of 20 μl 37% HCl. Tridecanoic acid in final concentration of 100 μM was added as an internal standard and the reaction mixture was extracted with 1 ml diethyl ether. After evaporation of the organic phase, the residue was dissolved in 40 μl *N,O*-bis(trimethylsilyl)trifluoroacetamide containing 1% trimethylchlorsilane and incubated at 75 °C for 30 min prior to analysis. 1 μl of each sample was analyzed by gas chromatography coupled to mass spectrometry (GC/MS) on a GC/MS QP-2010 instrument (Shimadzu, Japan) equipped with a FS-Supreme-5 column (30 m × 0.25 mm × 0.25 μm, Chromatographie Service GmbH, Langerwehe, Germany) and with helium as carrier gas. The injector and detector temperatures were set at 250 °C and 285 °C, respectively. The column temperature was set at 150 °C for 2 min, raised to 235 °C at a rate of 10 °C/min, and then to 300 °C at 30 °C/min. Reaction products were identified by their characteristic mass fragmentation patterns (Supplementary material, Fig. S1).

2.8. Determination of hydrogen peroxide formation

Reaction mixtures (0.8 ml) containing CacFld1, FdR, FdR/CacFld1, or FdR/CacFld1/CYP152A2 were prepared. The concentrations utilized were 0.25 μM CYP152A2, 1.25 μM FdR, 6.25 μM CacFld1. Myristic acid was added to a final concentration of 150 μM. The reaction was performed at 25 °C and started by adding 150 μM or 300 μM NADPH. NADPH consumption was monitored at 340 nm ($\varepsilon_{340} = 6.22$ mM^{-1} cm^{-1}). After 10 min, 0.5 ml were removed for the estimation of H_2O_2 formed by uncoupling. The remaining volume was used for GC/MS analysis. Samples were treated as described in Section 2.7.

H_2O_2 formed by uncoupling during NADPH oxidation was measured by the horseradish peroxidase/phenol/4-aminoantipyrine assay as described elsewhere [40]. The assay was carried out in 50 mM Tris–HCl pH 7.5 (final volume 0.8 ml), and contained 0.5 ml of the incubated reaction mixture as described above, 12.5 mM phenol, 1.25 mM 4-aminoantipyrine, and 0.1 mg/ml horseradish peroxidase. The absorbance at 510 nm (λ_{max} of the quinoneimine product) was recorded and the concentration of H_2O_2 in each reaction mixture was calculated from a calibration curve using known concentrations of

H_2O_2. In order to verify if the colour shift observed in the reaction could be attributed to H_2O_2, control samples containing catalase (600 U/ml) were assayed.

2.9. Steady-state kinetic parameters determination

Kinetic parameters for myristic acid conversion were determined by measuring the initial activities of CYP152A2 in reaction sets containing NADPH, FdR, CacFld1, catalase, and G6P/G6PDH at the concentrations mentioned in Section 2.7. Myristic acid concentrations ranged from 10 to 300 μM. Conversion was stopped after 15 min, followed by addition of 100 μM tridecanoic acid as internal standard. Sample treatment for GC/MS analysis was performed as described. Data were fit to the Michaelis–Menten equation by non-linear regression using "OriginPro 8G SR1" software (OriginLab Corporation, Northampton, USA) to obtain K_M and k_{cat}.

3. Results

3.1. Expression, purification and spectral characterization of flavoproteins

Two flavodoxins – CacFld1 and CacFld2 – and a flavodoxin reductase – CacFdR – from *C. acetobutylicum* were expressed in soluble form in *E. coli* BL21(DE3) and purified by immobilized metal affinity chromatography (IMAC). Flavodoxin yields were 2.8 mg L^{-1} CacFld1 and 4.4 mg L^{-1} CacFld2. Protein solutions containing purified flavodoxins were bright yellow, indicating that they were in their oxidized (quinone) forms. CacFld1 had two major absorption bands with maxima at 374 and 446 nm, with a shoulder on the longer wavelength band at 470 nm. CacFld2 had similar properties with maxima at 374 and 458 nm (Supplementary material, Fig. S2). The flavin of both flavodoxins was shown to be fully reduced by the chemical reductant sodium dithionite.

CacFdR from *C. acetobutylicum* was produced in a yield of 13 mg L^{-1}. Although the recombinant reductase from *C. acetobutylicum* could be expressed in soluble form in *E. coli*, the incorporation of FAD upon protein folding was not sufficient, which has been proven by spectral analysis (data not shown). Further protein downstream processes such as IMAC purification and dialysis increased the loss of the FAD cofactor. The addition of external FAD during expression and purification steps did not improve the properties of the flavodoxin reductase. A further attempt to obtain holoflavoprotein was the deflavination of CacFdR on a Ni-NTA column and its reconstitution by continuous circulation of free FAD [41]. Nevertheless, the cofactor failed to bind the apoprotein.

Therefore, since the endogenous flavodoxin and flavodoxin reductase system from *E. coli* (FdR) were found to support the activity of a variety of heterologously expressed P450s, such as CYP107H1 (P450 BioI) or CYP152A2 [12,16], the experiments on reduction of flavodoxins from *C. acetobutylicum* and on reconstitution of CYP152A2 activity were performed with FdR. FdR was expressed in *E. coli* BL21(DE3) in soluble form in a yield of 25 mg L^{-1} and purified. The spectral properties of oxidized FdR were identical to those reported previously [31].

3.2. Reduction of flavodoxins by FdR

Reduction of the flavodoxins through FdR was measured by mixing FdR and Fld in a ratio of 1:5, in order to enable the chromophore of the flavodoxins to be more easily recognized without any substantial masking by FdR. Absorbance changes on the key wavelengths 456 nm, 585 nm, and 640 nm after addition of NADPH were followed over time (Fig. 1). Upon addition of NADPH to the system with CacFld1, the absorbance at 456 nm dropped to 25% of its initial value. During this time, absorbance at 585 nm and 640 nm increased, demonstrating reduction of the flavin quinone to the neutral semiquinone form [42]. Afterwards the absorbance did not change, indicating that the semiquinone radical was not oxidized back to the quinone form.

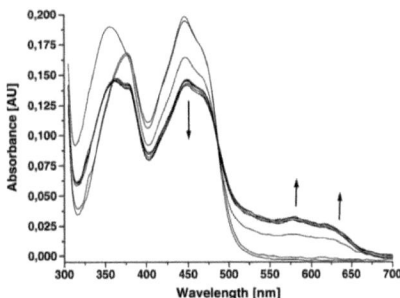

Fig. 1. UV-visible spectra monitoring electron transfer from FdR to CacFld1. Addition of NADPH causes the absorbance at 456 nm to drop by 25% after 1 min. The increase in absorbance at 585 nm and 640 nm is indicative of the reduction of the fully oxidized flavodoxin form to the neutral flavin semiquinone form.

In the experiment with the second flavodoxin CacFld2, no reduction of the flavin chromophore at 456 nm was observed, indicating that no electron transfer between FdR and the flavodoxin occurred. Due to this lack of interaction, the possible influence of the second flavodoxin on fatty acid hydroxylation activity by CYP152A2 could not be investigated further.

Furthermore, the role of CacFld1 in electron transfer from FdR to reduce cytochrome *c* was investigated. Purified FdR reduced cytochrome *c* with a K_M of 24 μM and a k_{cat} of 111 min^{-1} at 25 °C. These values are comparable to previously published data [31,43]. Addition of CacFld1 (1:5 FdR:Fld ratio) caused an increase in cytochrome *c* reduction rate from 21.1 to 56.2 μmol (μmol FdR)$^{-1}$ min^{-1}, which demonstrates that CacFld1 is able to transfer electrons to cytochrome *c* as was also observed for other flavodoxins [31,43].

3.3. Reduction of CYP152A2 heme-iron by FdR and CacFld1

Reconstitution of P450 activity requires electron delivery to the heme iron by auxiliary redox proteins. This reduction can be monitored by addition of carbon monoxide to the ferrous P450 (Fe^{2+}), which yields a stable complex with a characteristic absorbance at 450 nm [36,37]. Spectral scanning for the reaction of FdR and CacFld1 with CYP152A2 maintained under anaerobic conditions with N$_2$-flushing (see Section 2 for details) showed the appearance of the characteristic peak at 448 nm, indicative of P450 reduction and formation of the typical ferrous-carbon monoxide complex. In the system with both FdR and CacFld1, 18% reduction relative to dithionite-reduced P450 was achieved after 70 min (Fig. 2). Such a period of time was considered for monitoring since it was observed that the P450 heme-iron was slowly reduced in this system. Further, the GOD/CAT system was used to improve the anaerobic conditions (see Section 2 for details). In this system, glucose oxidation results in oxygen consumption by glucose oxidase and H_2O_2 formation, which is simultaneously scavenged by catalase to prevent enzyme inactivation. When both flavoproteins and CYP152A2 were maintained in this environment, 85% heme-iron reduction was reached. In addition it was observed that H_2O_2 alone was unable to reduce the P450 heme-iron (data not shown). These results clearly confirm that FdR and CacFld1 function as electron transfer partners of CYP152A2.

3.4. Reconstitution of CYP152A2 hydroxylase activity with FdR and CacFld1

Substrate conversion rates and conversion of myristic acid to α- and β-hydroxymyristic acid by CYP152A2 were measured in reconstituted enzyme systems containing CacFld1 and FdR, with and

Table 1
Myristic acid hydroxylation by CYP152A2 in reconstituted electron transfer chains.[a]

Electron transfer chain	Catalase (+), Cofactor Regeneration (−)		Catalase (+), Cofactor Regeneration (+)		Catalase (−), Cofactor Regeneration (−)		Catalase (−), Cofactor regeneration (+)	
	SCR[b]	Conversion [%][c]	SCR[b]	Conversion [%][c]	SCR[b]	Conversion [%][c]	SCR[b]	Conversion [%][c]
NADPH/FdR/CacFld1/P450	32	3.9	564	67.7	730	87.6	832	99.8
NADPH/FdR/P450	24	2.9	34	4.1	787	94.4	789	94.5
NADPH/CacFld1/P450	<6	<1	<6	<1	38	4.5	81	9.7
NADPH/P450	–	–	<6	<1	<6	<1	<6	<1

– not detected; conversion products below the detection limit (1 μM).
(−) absent; (+) present.
[a] The reported values are the averages of at least three measurements for which the standard deviations were less than 10% of the mean.
[b] Substrate conversion rates (SCR) are given in nmol substrate (nmol P450)$^{-1}$ h^{-1}.
[c] Quantitative analysis of conversions < 1% provided unreliable data.

without catalase. Both sets were assayed in the presence and absence of the G6P/G6PDH cofactor regenerating system. Conversion of myristic acid by CYP152A2 alone in the absence of redox proteins (background activity) was also determined. The results are summarized in Table 1.

The substrate conversion rates in the absence of catalase in systems containing CacFld1 and FdR (730 nmol substrate (nmol P450)$^{-1}$ h^{-1}) or containing FdR only (787 nmol substrate (nmol P450)$^{-1}$ h^{-1}) were similar. Furthermore, these sets were similar independently on cofactor regeneration. This suggests that without addition of catalase, hydroxylation was largely attributed to the peroxygenase activity of CYP152A2. Apparently electron uncoupling occurs, leading to oxygen reduction with formation of H_2O_2 and subsequent substrate oxidation via the peroxide-shunt. In the reaction set lacking FdR, the substrate conversion rate decreased significantly (38 nmol substrate (nmol P450)$^{-1}$ h^{-1}) in relation to the complete or flavodoxin-free sets, which implies that H_2O_2 generation mainly occurs when FdR is present.

Activity in the set NADPH/CacFld1/P450 without FdR is explained by generation of low H_2O_2 concentrations (Table 2). The increased conversion in this set without catalase but with cofactor regeneration (81 nmol substrate (nmol P450)$^{-1}$ h^{-1}) might be explained by the higher supply of electrons, used for O_2 reduction to H_2O_2.

When H_2O_2 was scavenged by catalase, substrate hydroxylation could be solely attributed to the monooxygenase activity of CYP152A2 in the presence of the redox proteins. Concerning the role of CacFld1, the complete reaction set with flavodoxin displayed 17-fold higher activity than without flavodoxin (Table 1; 68% versus 4% conversion, respectively), demonstrating that CacFld1 supports fatty acid hydroxylation by CYP152A2.

Interestingly, FdR alone is able to support P450 catalysis (as well as reduction of CYP152A2, Fig. 2B), although at a low extent. Substrate conversion in the set NADPH/FdR/P450 without cofactor regeneration upon catalase addition achieved 4%. To our knowledge, such interaction has not been described previously and should be studied in detail. In addition, as the substrate conversion rates in the set lacking FdR were < 1%, it is evident that the reductase is necessary for flavodoxin reduction to allow electron transfer to CYP152A2. Since no reduction of CacFld1 by NADPH in the absence of FdR was observed (data not shown), we suggest that substrate conversion in this case is identical to the background conversion observed with CYP152A2 alone in the absence of redox proteins upon cofactor regeneration (Table 1).

The hydroxylation activity in the sets containing FdR, CacFld1, and CYP152A2 was ~30% lower when catalase was present (564 nmol substrate (nmol P450)$^{-1}$ h^{-1}), than in the same set without catalase (832 nmol substrate (nmol P450)$^{-1}$ h^{-1}), indicating that a higher conversion is achieved as a sum of both peroxygenase and monooxygenase activity.

3.5. Determination of hydrogen peroxide formation

To elucidate the steps of the electron transfer chain by which H_2O_2 was formed, systems were set up containing either CacFld1 alone, FdR alone, FdR together with CacFld1, or FdR together with CacFld1 and CYP152A2. NADPH was given in a ratio 1:1 or 2:1 in relation to myristic acid in order to investigate if an excess of the electron donor resulted in a higher generation of H_2O_2. NADPH consumption and H_2O_2 generation were determined. The results are summarized in Table 2.

Concerning NADPH consumption, in systems containing FdR alone, similar amounts of NADPH were consumed after 10 min when using the NADPH/substrate ratios of 1:1 and 2:1 (Table 2). These data show that NADPH consumption by FdR alone proceeds at a constant rate in the absence of a flavodoxin. In contrast to these results, most NADPH (95–98%) was consumed in the two other systems adhering additionally CacFld1 and CYP152A2, even in cases where NADPH was in excess.

Concerning H_2O_2 formation, in the system containing CacFld1 alone, less than 2 μM hydrogen peroxide was detected in the incubation mixture with a 2:1 NADPH/substrate ratio. The highest concentration of H_2O_2 (90 μM) was observed in the system FdR-CacFld1 adhering a twofold excess of NADPH (Table 2). This suggests that electron uncoupling leading to H_2O_2 generation occurs mainly between FdR and CacFld1, and that more H_2O_2 is formed when more NADPH is available. As seen in Table 2, similar amounts of H_2O_2 were

Table 2
NADPH consumption, H_2O_2 formation, and substrate conversion in reconstituted electron transfer chains.[a]

Electron transfer chain	1:1 NADPH/substrate ratio (150 μM NADPH)				2:1 NADPH/substrate ratio (300 μM NADPH)			
	Consumed NADPH		Detected H_2O_2 [μM]	Substrate conversion [%]	Consumed NADPH		Detected H_2O_2 [μM]	Substrate conversion [%]
	[μM]	[%]			[μM]	[%]		
NADPH/CacFld1	–	–	–	–	–	–	1.9	–
NADPH/FdR	79	53	11.9	–	77	26	11.2	–
NADPH/FdR/CacFld1	143	95	59.9	–	294	98	90.0	–
NADPH/FdR/CacFld1/P450	145	97	3.4	83	286	95	35.8	85

– not detected; conversion products below the detection limit (1 μM).
[a] The reported values are the averages of at least three measurements for which the standard deviations were less than 10% of the mean.

Please cite this article as: S. Honda Malca, et al., Expression, purification and characterization of two Clostridium acetobutylicum flavodoxins: Potential electron transfer partners ..., Biochim. Biophys. Acta (2010), doi:10.1016/j.bbapap.2010.06.013

consumed by CYP152A2, independently on the NADPH concentration. Furthermore, in terms of substrate turnover, there was no significant difference between the conversion values with sufficient or excessive NADPH.

In order to verify efficient scavenging of the generated H_2O_2 by catalase, allowing exclusion of the peroxygenase activity of CYP152A2, catalase was added to the samples. In the presence of catalase almost no H_2O_2 (<1 μM) was estimated for all systems. In this case, conversion of myristic acid achieved 2.2% with 150 μM NADPH and 11.5% with 300 μM NADPH.

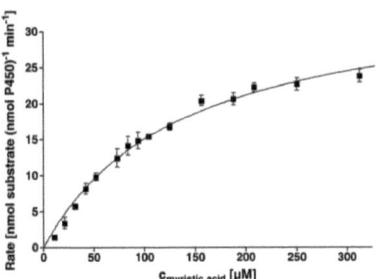

Fig. 3. Michaelis–Menten plot for myristic acid hydroxylation by CYP152A2.

3.6. Steady-state kinetic parameter determination for myristic acid hydroxylation by CYP152A2 with FdR and CacFld1

As CacFld1 was shown to interact with CYP152A2, kinetic constants for myristic acid conversion were determined in reconstituted systems containing FdR, CacFld1, CYP152A2, and catalase. The system exhibited a K_M for myristic acid of 137 μM and a k_{cat} of 36 min^{-1} (Fig. 3). The turnover rate of this system is considerably lower than the reported 200 min^{-1} with the same substrate when using H_2O_2; however, this activity was observed at a high concentrations of H_2O_2 (200 μM) only, which caused enzyme inactivation within 2–4 min [16].

4. Discussion

The purpose of this study was to identify physiological flavoproteins from *C. acetobutylicum* and to evaluate their functional role as potential redox partners of CYP152A2, a cytochrome P450 monooxygenase from the same microorganism. High oxidizing activities for myristic acid in the presence of both, H_2O_2 and NADPH-containing systems with *E. coli* flavodoxin reductase and *B. megaterium* CYP102A1-reductase described previously, indicated that CYP152A2 accepts H_2O_2 and molecular oxygen [16].

The genes encoding CacFld1, CacFld2, and CacFdR were selected as electron transfer candidates for CYP152A2, since others were predicted as multimeric, diverged, or disrupted [20]. From the three flavoproteins investigated, CacFdR could not be expressed and purified as a holoflavoprotein in sufficient yield; therefore, *E. coli* flavodoxin reductase (FdR) was employed instead.

The flavodoxin CacFld2 did not accept electrons from NADPH-reduced FdR. However, this flavodoxin cannot be excluded as a candidate redox partner for CYP152A2 in the presence of a suitable physiological flavodoxin reductase.

Given that electron transfer from reduced FdR was observed only for CacFld1, conversion of myristic acid by CYP152A2 was investigated with this flavodoxin only. Since H_2O_2 was demonstrated to be generated by electron uncoupling utilizing this reconstituted electron transfer chain, catalase was used as H_2O_2 scavenger in order to exclude the peroxygenase activity of CYP152A2. Indeed, in catalase-free systems substrate oxidation was largely attributed to the peroxygenase activity of CYP152A2, whilst the influence of CacFld1 on the monooxygenase activity of this enzyme became solely evident in the systems with catalase. These results, along with those providing evidence of reduction of CYP152A2 by reduced CacFld1, demonstrate that CacFld1 can serve as electron transfer protein for CYP152A2, since both proteins are originating from *C. acetobutylicum*.

The proposed electron transfer chain mechanism from NADPH to CYP152A2 via FdR and CacFld1 is summarized in Fig. 4. Electrons are

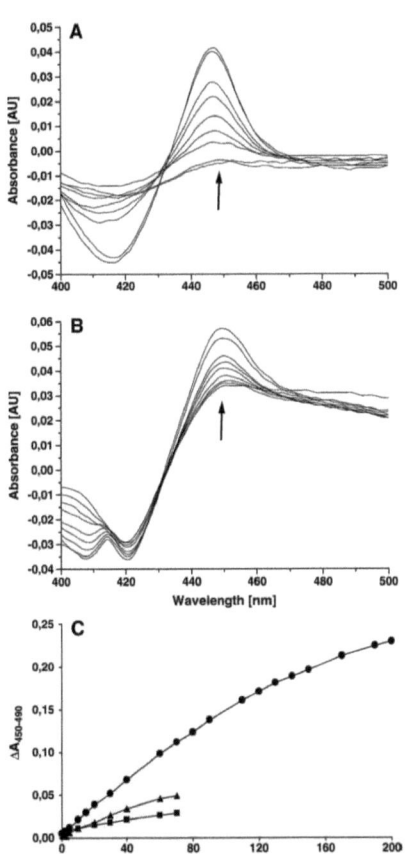

Fig. 2. CO-differential spectra of NADPH-reduced CYP152A2. (A) with FdR/CacFld1 and (B) with FdR only. The arrows indicate increasing peaks at 448 nm; measurements were performed every 8 min. (C) Absorbance difference vs. time: with FdR only (squares), with FdR/CacFld1 (triangles), and with FdR/CacFld1 in the presence of the GOD/CAT-oxygen scavenging system (circles).

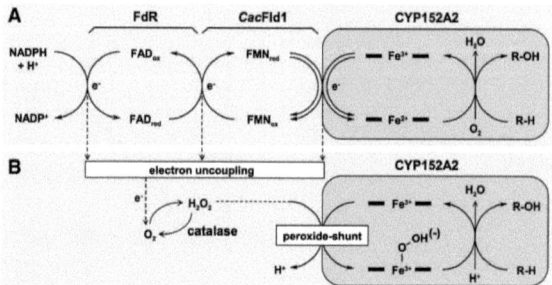

Fig. 4. Proposed electron transfer chain for CYP152A2. (A) Electrons are delivered from NADPH to the redox proteins FdR and CacFld1 and further to the heme group of CYP152A2. In the P450 catalytic site the reduced iron form can bind dioxygen, which results in substrate oxidation and the generation of one water molecule. (B) Alternatively, the production of H_2O_2 – caused by electron uncoupling from inefficient interactions between the redox proteins – can enter the catalytic cycle via the peroxide-shunt, in which ferric iron binds the hydroperoxide anion form. After a series of events (not detailed here), the substrate is oxygenated. H_2O_2 can be scavenged by catalase to prevent product formation via the peroxide-shunt. Abbreviations: ox: oxidized, red: reduced.

delivered from NADPH to the redox proteins FdR and CacFld1 and then to the heme group of CYP152A2. In the P450 catalytic site the reduced ferric form (ferrous) can bind dioxygen, which results in substrate oxidation and the generation of one water molecule. Alternatively, the production of H_2O_2 – caused by electron uncoupling from inefficient interactions between the redox partners – can enter the catalytic cycle via the peroxide-shunt, in which ferric iron binds the hydroperoxide anion form. This will also lead to substrate oxidation via a series of events (not detailed here).

The fact that only partial reduction of CacFld1 (25%) was achieved by FdR reflects the inefficient interaction between the two flavoproteins, an expected phenomenon since FdR is a non-natural electron transfer partner for CacFld1. Without a physiological reductase from C. acetobutylicum at hand, no reliable conclusion can be done, whether peroxygenase activity (for scavenging of H_2O_2) and monooxygenase activity (for detoxification of O_2) exist independently of each other, because uncoupling may be reduced or negligible in the physiologically relevant electron transport pathway.

In summary, CYP152A2 might represent the next P450 beside P450cin belonging to the class III system. However, further investigations of CYP152A2 with a physiological reductase will provide more insights into the mechanism of electron transfer in this system, as well into its physiological function in C. acetobutylicum.

Acknowledgments

SHM, MG and VBU acknowledge the support of this work by Deutsche Forschungsgemeinschaft (SFB706) and the Ministerium für Wissenschaft, Forschung und Kunst des Landes Baden-Württemberg.

Appendix A. Supplementary material

Supplementary data associated with this article can be found, in the online version, at doi:10.1016/j.bbapap.2010.06.013.

References

[1] R. Bernhardt, Cytochromes P450 as versatile biocatalysts, J. Biotechnol. 124 (2006) 128–145.
[2] D.R. Nelson, Cytochrome P450 nomenclature, 2004, Methods Mol. Biol. 320 (2006) 1–10.
[3] F. Hannemann, A. Bichet, K.M. Ewen, R. Bernhardt, Cytochrome P450 systems – biological variations of electron transport chains, Biochim. Biophys. Acta 1770 (2007) 330–344.
[4] K.J. McLean, M. Sabri, K.R. Marshall, R.J. Lawson, D.G. Lewis, D. Clift, P.R. Balding, A.J. Dunford, A.J. Warman, J.P. McVey, A.M. Quinn, M.J. Sutcliffe, N.S. Scrutton, A.W. Munro, Biodiversity of cytochrome P450 redox systems, Biochem. Soc. Trans. 33 (2005) 796–801.
[5] D.B. Hawkes, G.W. Adams, A.L. Burlingame, P.R. Ortiz de Montellano, J.J. De Voss, Cytochrome P450(cin) (CYP176A), isolation, expression, and characterization, J. Biol. Chem. 277 (2002) 27725–27732.
[6] R.M. Smillie, Isolation of two proteins with chloroplast ferredoxin activity from a blue-green alga, Biochem. Biophys. Res. Commun. 20 (1965) 621–629.
[7] E. Knight Jr., A.J. D'Eustachio, R.W. Hardy, Flavodoxin: a flavoprotein with ferredoxin activity from Clostridium pasteurianum, Biochim. Biophys. Acta. 113 (1966) 626–628.
[8] J. Sancho, Flavodoxins: sequence, folding, binding, function and beyond, Cell. Mol. Life Sci. 63 (2006) 855–864.
[9] K. Fujii, J.H. Galivan, F.M. Huennekens, Activation of methionine synthase: further characterization of flavoprotein system, Arch. Biochem. Biophys. 178 (1977) 662–670.
[10] H.J. Barnes, M.P. Arlotto, M.R. Waterman, Expression and enzymatic activity of recombinant cytochrome P450 17 alpha-hydroxylase in Escherichia coli, Proc. Natl. Acad. Sci. U. S. A. 88 (1991) 5597–5601.
[11] C.M. Jenkins, M.R. Waterman, Flavodoxin and NADPH-flavodoxin reductase from Escherichia coli support bovine cytochrome P450c17 hydroxylase activities, J. Biol. Chem. 269 (1994) 27401–27408.
[12] R.J. Lawson, C. von Wachenfeldt, I. Haq, J. Perkins, A.W. Munro, Expression and characterization of the two flavodoxin proteins of Bacillus subtilis, YkuN and YkuP: biophysical properties and interactions with cytochrome P450 BioI, Biochemistry 43 (2004) 12390–12409.
[13] M. Girhard, T. Klaus, Y. Khatri, R. Bernhardt, V.B. Urlacher, Characterization of the versatile monooxygenase CYP109B1 from Bacillus subtilis, Appl. Microbiol. Biotechnol. 87 (2) (2010) 595–607.
[14] M. Girhard, K. Machida, M. Itoh, R.D. Schmid, A. Arisawa, V.B. Urlacher, Regioselective biooxidation of (+)-valencene by recombinant E. coli expressing CYP109B1 from Bacillus subtilis in a two-liquid-phase system, Microb. Cell. Fact. 8 (2009) 36.
[15] A.J. Green, A.W. Munro, M.R. Cheesman, G.A. Reid, C. von Wachenfeldt, S.K. Chapman, Expression, purification and characterisation of a Bacillus subtilis ferredoxin: a potential electron transfer donor to cytochrome P450 BioI, J. Inorg. Biochem. 93 (2003) 92–99.
[16] M. Girhard, S. Schuster, M. Dietrich, P. Durre, V.B. Urlacher, Cytochrome P450 monooxygenase from Clostridium acetobutylicum: a new alpha-fatty acid hydroxylase, Biochem. Biophys. Res. Commun. 362 (2007) 114–119.
[17] I. Matsunaga, N. Yokotani, O. Gotoh, E. Kusunose, M. Yamada, K. Ichihara, Molecular cloning and expression of fatty acid alpha-hydroxylase from Sphingomonas paucimobilis, J. Biol. Chem. 272 (1997) 23592–23596.
[18] I. Matsunaga, A. Ueda, N. Fujiwara, T. Sumimoto, K. Ichihara, Characterization of the ybdT gene product of Bacillus subtilis: novel fatty acid beta-hydroxylating cytochrome P450, Lipids 34 (1999) 841–846.
[19] S. Keis, C.F. Bennett, V.K. Ward, D.T. Jones, Taxonomy and phylogeny of industrial solvent-producing clostridia, Int. J. Syst. Bacteriol. 45 (1995) 693–705.
[20] J. Nölling, G. Breton, M.V. Omelchenko, K.S. Makarova, Q. Zeng, R. Gibson, H.M. Lee, J. Dubois, D. Qiu, J. Hitti, Y.I. Wolf, R.L. Tatusov, F. Sabathe, L. Doucette-Stamm, P. Soucaille, M.J. Daly, G.N. Bennett, E.V. Koonin, D.R. Smith, Genome sequence and comparative analysis of the solvent-producing bacterium Clostridium acetobutylicum, J. Bacteriol. 183 (2001) 4823–4838.
[21] H. Seedorf, H.W. Fricke, B. Veith, H. Brüggemann, H. Liesegang, A. Strittmatter, M. Miethke, W. Buckel, J. Hinderberger, F. Li, C. Hagemeier, R.K. Thauer, G. Gottschalk, The genome of Clostridium kluyveri, a strict anaerobe with unique metabolic features, Proc. Natl. Acad. Sci. U. S. A. 105 (2008) 2128–2133.

[22] S. Kawasaki, Y. Watamura, M. Ono, T. Watanabe, K. Takeda, Y. Niimura, Adaptive responses to oxygen stress in obligatory anaerobes *Clostridium acetobutylicum* and *Clostridium aminovalericum*, Appl. Environ. Microbiol. 71 (2005) 8442–8450.
[23] V. Briolat, G. Reysset, Identification of the *Clostridium perfringens* genes involved in the adaptive response to oxidative stress, J. Bacteriol. 184 (2002) 2333–2343.
[24] H. Brüggemann, R. Bauer, S. Raffestin, G. Gottschalk, Characterization of a heme oxygenase of *Clostridium tetani* and its possible role in oxygen tolerance, Arch. Microbiol. 182 (2004) 259–263.
[25] F. Hillmann, C. Doring, O. Riebe, A. Ehrenreich, R.J. Fischer, H. Bahl, The role of PerR in O2-affected gene expression of *Clostridium acetobutylicum*, J. Bacteriol. 191 (2009) 6082–6093.
[26] F. Hillmann, R.J. Fischer, F. Saint-Prix, L. Girbal, H. Bahl, PerR acts as a switch for oxygen tolerance in the strict anaerobe *Clostridium acetobutylicum*, Mol. Microbiol. 68 (2008) 848–860.
[27] F. Hillmann, O. Riebe, R.J. Fischer, A. Mot, J.D. Caranto, D.M. Kurtz Jr., H. Bahl, Reductive dioxygen scavenging by flavo-diiron proteins of *Clostridium acetobutylicum*, FEBS Lett. 583 (2009) 241–245.
[28] D. Jean, V. Briolat, G. Reysset, Oxidative stress response in *Clostridium perfringens*, Microbiology 150 (2004) 1649–1659.
[29] M. Demuez, L. Cournac, O. Guerrini, P. Soucaille, L. Girbal, Complete activity profile of *Clostridium acetobutylicum* [FeFe]-hydrogenase and kinetic parameters for endogenous redox partners, FEMS Microbiol. Lett. 275 (2007) 113–121.
[30] O. Guerrini, B. Burlat, C. Leger, B. Guigliarelli, P. Soucaille, L. Girbal, Characterization of two 2[4Fe4S] ferredoxins from *Clostridium acetobutylicum*, Curr. Microbiol. 56 (2008) 261–267.
[31] C.M. Jenkins, M.R. Waterman, NADPH-flavodoxin reductase and flavodoxin from *Escherichia coli*: characteristics as a soluble microsomal P450 reductase, J. Biol. Chem. 269 (1994) 6106–6113.
[32] R.P. Swenson, G.D. Krey, Site-directed mutagenesis of tyrosine-98 in the flavodoxin from *Desulfovibrio vulgaris* (Hildenborough): regulation of oxidation-reduction properties of the bound FMN cofactor by aromatic, solvent, and electrostatic interactions, Biochemistry 33 (1994) 8505–8514.
[33] L. Liu, R.D. Schmid, V.B. Urlacher, Cloning, expression, and characterization of a self-sufficient cytochrome P450 monooxygenase from *Rhodococcus ruber* DSM 44319, Appl. Microbiol. Biotechnol. 72 (2006) 876–882.
[34] K. Fujii, F.M. Huennekens, Activation of methionine synthetase by a reduced triphosphopyridine nucleotide-dependent flavoprotein system, J. Biol. Chem. 249 (1974) 6745–6753.
[35] B. van Gelder, E.C. Slater, The extinction coefficient of cytochrome c, Biochim. Biophys. Acta 58 (1962) 593–595.
[36] T. Omura, R. Sato, The carbon monoxide-binding pigment of liver microsomes. I. Evidence for its hemoprotein nature, J. Biol. Chem. 239 (1964) 2370–2378.
[37] T. Omura, R. Sato, The carbon monoxide-binding pigment of liver microsomes. II. Solubilization, purification, and properties, J. Biol. Chem. 239 (1964) 2379–2385.
[38] V.V. Shumyantseva, V. Uvarov, O.E. Byakova, A.I. Archakov, Semisynthetic flavocytochromes based on cytochrome P450 2B4: reductase and oxygenase activities, Arch. Biochem. Biophys. 354 (1998) 133–138.
[39] T. Matsubara, J. Baron, L.L. Peterson, J.A. Peterson, NADPH-cytochrome P450 reductase, Arch. Biochem. Biophys. 172 (1976) 463–469.
[40] F. Xu, S.G. Bell, Z. Rao, L.L. Wong, Structure-activity correlations in pentachlorobenzene oxidation by engineered cytochrome P450cam, Protein Eng. Des. Sel. 20 (2007) 473–480.
[41] M.H. Hefti, F.J. Milder, S. Boeren, J. Vervoort, W.J. van Berkel, A His-tag based immobilization method for the preparation and reconstitution of apoflavoproteins, Biochim. Biophys. Acta 1619 (2003) 139–143.
[42] F. Müller, M. Brustlein, P. Hemmerich, V. Massey, W.H. Walker, Light-absorption studies on neutral flavin radicals, Eur. J. Biochem. 25 (1972) 573–580.
[43] L. McIver, C. Leadbeater, D.J. Campopiano, R.L. Baxter, S.N. Daff, S.K. Chapman, A.W. Munro, Characterisation of flavodoxin NADP+ oxidoreductase from *Escherichia coli*: key components of electron transfer in *Escherichia coli*, Eur. J. Biochem. 257 (1998) 577–585.

2.2.2.1. Supplementary Material

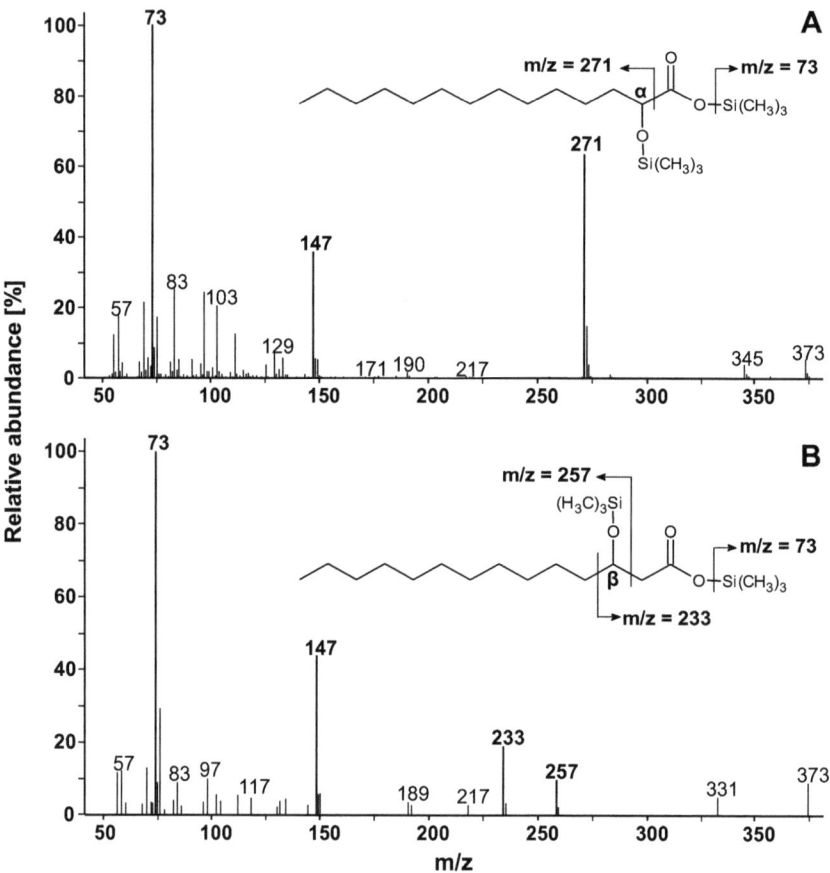

Figure 2-19 (supplementary material, Fig. S1): Mass spectra of trimethylsiloxyl (TMS) esters identified as α-hydroxy myristic acid (A) and β-hydroxy myristic acid (B). The common fragment ion at m/z = 73 results from the cleavage of the TMS ester group. The fragment ion at m/z = 147, which is characteristic of polysylated compounds, involves the loss of a methyl radical from one silyl group and its interaction with another TMS ester group. The fragment ions at m/z = 233, 257 and 271 represent typical fragmentations indicated in the corresponding structures.

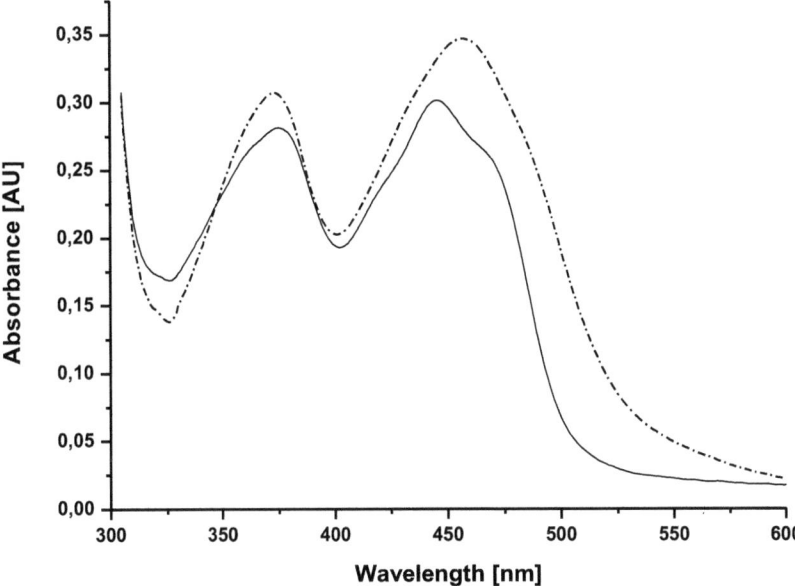

Figure 2-20 (supplementary material, Fig. S2): Spectral properties of oxidized *Cac*Fld1 (solid line) and oxidized *Cac*Fld2 (dashed-dotted line).

3. Discussion and Outlook

3.1. Cytochrome P450 monooxygenases for biocatalysis

3.1.1. Screening, protein expression and purification

Within this study a screening for new cytochrome P450 monooxygenases capable of producing fine chemicals like flavors and fragrances was carried out [140]. A recombinant *E. coli* library expressing 242 bacterial and fungal P450s that cover 89 families and 149 subfamilies was utilized [141]. Twenty doubtless hits were recognized that demonstrated activity against at least one of four substrates tested, namely geraniol, nerol, α-pinene and (+)-valencene (M. Girhard, unpublished data). Compared to the size of the library, however, the number of hits was lower than expected, since for example in a screening for oxidation of the steroid testosterone utilizing this library 35 hits could be recognized [142]. Only two P450s were identified capable of (+)-valencene oxidation: P450$_{MoxA}$ (CYP105) from *Nonomuraea recticatena* [143] and CYP109B1 from *B. subtilis*, but only CYP109B1 demonstrated the desired selectivity for allylic oxidation [140].

The second P450 of interest – CYP152A2 from *C. acetobutylicum* [144] – was chosen for two reasons: (i) it is the only known P450 found in an anaerobe bacterium and (ii) from the *C. acetobutylicum* genome sequencing project CYP152A2 was predicted to belong to the group of natural peroxygenases [145]. These P450s are interesting for biocatalysis, because they do not necessarily relay on external redox partners. For P450$_{Bsβ}$ (CYP152A1) – a peroxygenase from *B. subtilis* with 57% amino acid identity to CYP152A2 [146] – it was even stated that it is unable to except electrons via redox proteins [147]. Instead H_2O_2 is utilized for catalysis via the "peroxide shunt" [69]. Both, CYP152A2 and P450$_{Bsβ}$ show a unique substrate-enzyme interaction for oxidation of saturated fatty acids at the α- or β-carbon atom. A conserved arginine residue close to the heme iron at the distal side of the heme molecule (Arg^{242} in P450$_{Bsβ}$) can interact with the carboxylic group of the substrate, which determines the regioselectivity [147]. CYP152A2 could therefore represent a very useful biocatalyst for production of α- and β-hydroxylated fatty acids. Further, since it was shown within this study that CYP152A2 does not only function as peroxygenase, but can also act as monooxygenase accepting electrons from several redox proteins, this enzyme represents an interesting model P450 for detailed investigations on activity reconstitution.

CYP109B1 and CYP152A2 were cloned, expressed in *E. coli* and purified. pET-28a(+) expression vectors allowing high levels of protein expression in combination with *E. coli* strain BL21(DE3) were utilized for expression of both P450s. Expression levels of up to 120 mg l^{-1} for CYP109B1 and 21 mg l^{-1} for CYP152A2 were reached, which were purified by immobilized metal affinity chromatography (IMAC) to ~ 90-95% purity.

3.1.2. Substrate and product spectra

Binding and oxidation of various chemically different substances by CYP109B1 and CYP152A2 was investigated [144, 148]. The products of substrate oxidation were identified by means of gas chromatography coupled with mass spectrometry (GC/MS).

3.1.2.1. Substrate and product spectrum of CYP109B1

CYP109B1 turned out to be a versatile monooxygenase accepting a broad range of chemically different substrates. Generally, "linear" molecules (saturated fatty acids and their methyl- and ethylesters, unsaturated fatty acids, or *n*-alcohols) were oxidized unselectively resulting in more than three products, if saturated fatty acids and primary *n*-alcohols were oxidized, and even more than 10 products (most of which could not be identified) in the case of unsaturated fatty acids. Interestingly, though the oxidation proceeds unselectively, there is a slight preference for oxidation of carbon atoms C11 and C12 counted from the carboxylic group of (saturated) fatty acids. Such regioselectivity is unusual for fatty acid hydroxylating monooxygenases. The best examined enzyme in this respect is the self-sufficient monooxygenase CYP102A1 (P450$_{BM3}$) from *B. megaterium* [149]. The wildtype enzyme shows exclusive hydroxylation of the carbon atoms ω_{-1}, ω_{-2} and ω_{-3} preferring the ω_{-2} position [78]. Binding of fatty acids by CYP102A1 is arranged in a way that the carboxylic group is stabilized by amino acids at the entrance of the substrate channel (e.g. Arg^{47}), with the hydrophobic alkyl chain pointing towards the heme. It is possible that fatty acid binding by CYP109B1 proceeds in a way that the carboxylic group is anchored in a fixed distance from the heme resulting in the described preference for positions C11 and C12.

Remarkably, CYP109B1 shows high regioselectivity of oxidation at allylic carbon atoms of the cyclic sesquiterpene (+)-valencene and the norisoprenoids *α*- and *β*-ionone. In the case of (+)-valencene 65% (of the total products) of the desired allylic products nootkatol and (+)-nootkatone were received. The regioselectivity for *α*- and *β*-ionone was even 100% resulting in 3-hydroxy-α-ionone and 4-hydroxy-β-ionone exclusively. 4-hydroxy-β-ionone is an important intermediate for the synthesis of carotenoids (widely used as dietary supplement or as food coloring [150-151]) and also of deoxyabscisic acid, a synthetic analogue of the phytohormone abscisic acid (used as insect repellant [152]). In summary, CYP109B1 seems to be an attractive candidate for selective oxidation of terpenes, generating valuable compounds for flavor and fragrance industries.

CYP109D1 from *S. cellulosum* belonging to the same P450 family (CYP109) demonstrated identical regioselectivity and product specificity for ionone oxidation [153]. A sequence alignment using BLASTP 2.2.24 software [154] showed that the amino acid sequence of CYP109D1 has a similarity of 54% to CYP109B1, which is only 1% below the requirement for P450s belonging to the same subfamily (55%, see chapter 1.2.1). Since CYP109D1 is also capable of oxidizing saturated fatty acids (Y. Kathri and M. Girhard, personal communication and unpublished data), it is likely that this enzyme has a substrate spectrum similar to CYP109B1 and hence CYP109D1 represents another attractive biocatalyst with potential for biotechnological application.

The reason for the high regioselectivity of allylic oxidations by CYP109B1 and CYP109D1 is unclear at present, but might be caused by electronic activation of the secondary allylic carbon atoms: In the case of α-ionone, the carbon atom C3 next to the double bond between C4=C5 is most susceptible to an oxidative attack, while for β-ionone this holds true for C4 next to the C5=C6 double bond. 100% regioselectivity for β-ionone oxidation has to the best knowledge so far been described for mutated variants of P450$_{BM3}$ only, for example the triple-mutant A74E F87V P368S [155], but never for the wildtype enzyme or any other P450. Further, since a racemic mixture of α-ionone (6S- and 6R-α-ionone) was employed and conversion by CYP109B1 was higher than 50%, obviously both enantiomers were accepted as substrates. The non-enantioselective hydroxylation of racemic α-ionone at carbon atom C3 should result in a mixture of four diastereoisomers and consequently two isomers would be detected by achiral GC analysis (while four products would be detected by chiral analysis) [156]. The observed single peak observed during achiral GC analysis therefore might indicate that hydroxylation at C3 by CYP109B1 proceeds enantioselective, which would lead either to the 3S,6S- and 3R,6R-α-ionone enantiomers, or the 3R,6S- and 3S,6R-α-ionone enantiomers. No further work was carried out within this study to determine the exact R- and S-configurations of the products.

3.1.2.2. Substrate and product spectrum of CYP152A2

In contrast to the broad substrate spectra of CYP109 enzymes, the substrate spectrum of CYP152A2 seems to be limited to fatty acids and is therefore rather narrow. The regioselectivity for the oxidation of fatty acids, in contrast, is high leading to α- and β-hydroxylated products exclusively (with preference for the α-position). Such products can for example be utilized as precursors for synthesis of antibiotic compounds like surfactin [157-158]. Interestingly, a recent publication on P450$_{Bsβ}$ (CYP152A1, which was also examined in the course of CYP152A2-characterization) describes that the substrate spectrum of this enzyme could be extended towards cyclic substrates by applying a so-called "decoy-molecule", namely an alkyl-chain carboxylic acid shorter than ten carbon-atoms long [159]. P450$_{Bsβ}$ does not show catalytic activity towards such carboxylic acids, but their addition and binding in the substrate binding pocket allows the P450 enantioselective epoxidation of styrene, hydroxylation of ethylbenzene and oxidation of guaiacol. The reason for these novel activities was clarified by resolving the crystal structure of the decoy-molecule-bound form indicating that the carboxylic group of the decoy-molecule serves as general acid-base catalyst and that the role of the acid-base function is satisfied by the carboxylic group of these molecules [146]. Since CYP152A2 does also not show activity towards carboxylic acids shorter than ten carbon-atoms, a similar effect might be expected for application of decoy-molecules with CYP152A2 leading to new exciting perspectives for this biocatalyst.

Since wildtype enzymes were employed in all experiments described, the activity of the utilized P450s towards the described substrates (which is rather low at present) might be enhanced by several methods of protein design, like random mutagenesis or rational design. Numerous examples exist in literature and have been summarized in various reviews, where the activities and/or regioselectivities of P450s were improved dramatically by such means [160-164].

3.1.3. Activity reconstitution

The overall performance of a P450 catalyzed reaction does not only depend on the activity towards the substrate of interest of the P450 itself, but is dependent on a variety of other factors as well. One of the most important bottlenecks for biocatalysis with P450s is represented by electron transfer to the heme group via external redox partners, which is often considered as a rate limiting step, especially if non-physiological redox proteins are used to reconstitute the oxidizing activity [138, 165-166]. Inefficient electron transfer leads to uncoupling reactions meaning that the consumption of NAD(P)H proceeds without substrate oxidation. Hydrogen peroxide is formed under these conditions, which can lead to enzyme inactivation. This phenomenon was also observed within this study for CYP152A2, if high concentrations of hydrogen peroxide were used being necessary for efficient catalytic turnover rates.

Due to the observed high extend of uncoupling between non-physiological redox proteins, a search for possible physiological redox partners for CYP109B1 and CYP152A2 has become an important task within this study [148, 167]. Since the flavodoxins YkuN and YkuP from *B. subtilis* have already been described in literature to support activity of P450$_{Biol}$ (see chapter 1.3.2.3) [108], they were chosen as candidates for activity reconstitution of CYP109B1. Unfortunately a suitable candidate reductase from *B. subtilis* was not found when the complete annotated genome sequence of strain 168 [168] was analyzed. When searching the genome sequence of *C. acetobutylicum* ATCC 824 [145] for candidate redox partners for CYP152A2, however, a potential reductase annotated as "NADH-dependent flavine oxidoreductase" (GenBank CAC0196) could be identified. The gene was cloned, but the enzyme repeatedly failed to be expressed in functional form in *E. coli*. Therefore, flavodoxin reductase from *E. coli* (FdR, see chapter 1.3.2.1) was used to support both P450s.

For CYP109B1 none of the reconstituted redox systems employed reached 100% coupling between NAD(P)H consumption and substrate oxidation, which was obviously due to the fact that non-physiological redox partners had to be used. This hypothesis is supported by results gained for CYP152A2 demonstrating that highest hydrogen peroxide formation occurs between the redox proteins FdR from *E. coli* and the flavodoxin *Cac*Fld1 from *C. acetobutylicum*.

Interestingly, bovine truncated AdR and Adx were demonstrated to be efficient electron donors for CYP109B1 and CYP109D1, which has further been reported in literature for another bacterial P450, namely CYP106A2 from *B. megaterium* [169-170]. This is somehow surprising, since for example Pdx is not able to efficiently support the activity of CYP111 from *P. putida* [171], whereas P450$_{cam}$ (CYP101) from the same organism seems to function with Pdx exclusively [128]. Such behavior is unpredictable, and the explanation thereof is unclear at present and requires further detailed investigations.

Future investigations on electron transport systems for CYP109B1 and CYP152A2 should focus on the implementation of physiological reductases, which will possibly result in a reduction of uncoupling and consequently lead to higher P450 activities. This was demonstrated for example for CYP175A1 from the thermophilic bacterium *Thermus thermophilus* HB27, where the oxidation of the natural substrate β-carotene to β-cryptoxanthin could be enhanced 80-fold from 0.23 nmol (nmol P450)$^{-1}$ min^{-1} with artificial PdR-Pdx [138] to 18.3 nmol (nmol P450)$^{-1}$ min^{-1} by identification and application of a complete physiological electron transfer chain [172]. Further, ferredoxins like the [4S-4Fe] ferredoxin from *B. subtilis* [109] or other flavodoxins from both organisms could be tested for their ability to support activity of CYP109B1 and CYP152A2, respectively.

3.2. Whole-cell biocatalysis

CYP109B1 is capable of (+)-valencene oxidation to yield nootkatol, which can be further oxidized to the high-priced flavoring (+)-nootkatone (see chapter 2.1.2) [16]. A whole-cell process with recombinant *E. coli* strain BL21(DE3) expressing CYP109B1 together with PdR and Pdx (localized on a single plasmid) was developed [140]. Protein expression was under control of the T7-promoter working essentially as described in chapter 3.1.1, except that a protein-tag was not included, since protein purification was not necessary in this case.

When the biotransformation was carried out in an aqueous phase, nootkatol and (+)-nootkatone accounted for 65% of the total products. The other 35% of the total products were undesired byproducts arising from overoxidation of the monooxygenated products by CYP109B1. In order to protect the primary oxidation products from overoxidation, aqueous-organic two-liquid-phase systems with water immiscible organic solvents (*n*-octane, isooctane, dodecane, hexadecane) were set up and analyzed in terms of biocompatibility, substrate and product distribution, and volumetric productivities. The best result was achieved with a system comprising 10% dodecane. This solvent was benign for *E. coli* and demonstrated suitable partitioning of the substrate and the monooxygenated products. Under optimized conditions, less than 7% of undesired multioxygenated products were received and up to 15 mg l^{-1} h^{-1} nootkatol and nootkatone could be produced. This amount is much higher than previously reported values achieved with plant cells (~ 250 mg l^{-1} within 18 d for *Chlorella fusca* [173]; 600 mg l^{-1} within 20 d for *Gynostemma pentaphyllum* [174]) or non-recombinant bacteria (50 mg l^{-1} within 5 d for

Rhodococcus spec. [175]), and comparable to recent data reported for lyophilized cells of the fungi *Pleurotus sapidus* (~ 13.5 mg l^{-1} h^{-1}) [176].

The application of the developed two-phase system utilizing recombinant CYP109B1-expressing whole-cells therefore opens up new perspectives for biotechnological production of flavors and fragrances.

4. References

1. Schmid A, Hollmann F, Park JB, Buhler B: The use of enzymes in the chemical industry in Europe. *Curr Opin Biotechnol* 2002, 13:359-366.
2. Krings U, Berger RG: Biotechnological production of flavours and fragrances. *Appl Microbiol Biotechnol* 1998, 49:1-8.
3. Schmid A, Dordick JS, Hauer B, Kiener A, Wubbolts M, Witholt B: Industrial biocatalysis today and tomorrow. *Nature* 2001, 409:258-268.
4. Kaplan DL, Dordick J, Gross RA, Swift G: Enzymes in polymer science: An introduction. *Enzymes in Polymer Synthesis* 1998, 684:2-16.
5. Nagasawa T, Yamada H: Application of Nitrile Converting Enzymes for the Production of Useful Compounds. *Pure Appl Chem* 1990, 62:1441-1444.
6. Kirby J, Keasling JD: Biosynthesis of plant isoprenoids: perspectives for microbial engineering. *Annu Rev Plant Biol* 2009, 60:335-355.
7. Duetz WA, Bouwmeester H, van Beilen JB, Witholt B: Biotransformation of limonene by bacteria, fungi, yeasts, and plants. *Appl Microbiol Biotechnol* 2003, 61:269-277.
8. Schrader J: Microbial flavor production. In *Flavours and fragrances: Chemistry, Bioprocessing and Sustainability.* Edited by Berger RG. Berlin: Springer; 2007: 507-574.
9. Schrader J, Berger RG: Biotechnological production of terpenoid flavor and fragrance compounds. In *Biotechnology.* Edited by Rehm HJ, Reed G. Weinheim: Wiley-VCH; 2001: 373-422.
10. Dordick JS, Khmelnitsky YL, Sergeeva MV: The evolution of biotransformation technologies. *Curr Opin Microbiol* 1998, 1:311-318.
11. Maimone TJ, Baran PS: Modern synthetic efforts toward biologically active terpenes. *Nat Chem Biol* 2007, 3:396-407.
12. Serra S, Fuganti C, Brenna E: Biocatalytic preparation of natural flavours and fragrances. *Trends Biotechnol* 2005, 23:193-198.
13. Priefert H, Rabenhorst J, Steinbuchel A: Biotechnological production of vanillin. *Appl Microbiol Biotechnol* 2001, 56:296-314.
14. Dhavalikar RS, Bhattacharyya PK: Microbiological transformations of terpenes. 8. Fermentation of limonene by a soil pseudomonad. *Indian J Biochem* 1966, 3:144-157.
15. Brenna E, Fuganti C, Serra S: Enantioselective perception of chiral odorants. *Tetrahedron-Asymmetry* 2003, 14:1-42.

16. Fraatz MA, Berger RG, Zorn H: Nootkatone - a biotechnological challenge. *Appl Microbiol Biotechnol* 2009, 83:35-41.
17. Nelson DR: Cytochrome P450 nomenclature, 2004. *Methods Mol Biol* 2006, 320:1-10.
18. Garfinkel D: Studies on pig liver microsomes. I. Enzymic and pigment composition of different microsomal fractions. *Arch Biochem Biophys* 1958, 77:493-509.
19. Klingenberg M: Pigments of rat liver microsomes. *Arch Biochem Biophys* 1958, 75:376-386.
20. Nebert DW, Adesnik M, Coon MJ, Estabrook RW, Gonzalez FJ, Guengerich FP, Gunsalus IC, Johnson EF, Kemper B, Levin W, et al.: The P450 gene superfamily: recommended nomenclature. *DNA* 1987, 6:1-11.
21. Nebert DW, Nelson DR, Adesnik M, Coon MJ, Estabrook RW, Gonzalez FJ, Guengerich FP, Gunsalus IC, Johnson EF, Kemper B, et al.: The P450 superfamily: updated listing of all genes and recommended nomenclature for the chromosomal loci. *DNA* 1989, 8:1-13.
22. Nebert DW, Nelson DR, Coon MJ, Estabrook RW, Feyereisen R, Fujii-Kuriyama Y, Gonzalez FJ, Guengerich FP, Gunsalus IC, Johnson EF, et al.: The P450 superfamily: update on new sequences, gene mapping, and recommended nomenclature. *DNA Cell Biol* 1991, 10:1-14.
23. Nelson DR, Kamataki T, Waxman DJ, Guengerich FP, Estabrook RW, Feyereisen R, Gonzalez FJ, Coon MJ, Gunsalus IC, Gotoh O, et al.: The P450 superfamily: update on new sequences, gene mapping, accession numbers, early trivial names of enzymes, and nomenclature. *DNA Cell Biol* 1993, 12:1-51.
24. Nelson DR: Progress in tracing the evolutionary paths of cytochrome P450. *Biochim Biophys Acta* 2010, doi:10.1016/j.bbapap.2010.08.008.
25. Nelson DR, Koymans L, Kamataki T, Stegeman JJ, Feyereisen R, Waxman DJ, Waterman MR, Gotoh O, Coon MJ, Estabrook RW, et al: P450 superfamily: update on new sequences, gene mapping, accession numbers and nomenclature. *Pharmacogenetics* 1996, 6:1-42.
26. Fischer M, Knoll M, Sirim D, Wagner F, Funke S, Pleiss J: The Cytochrome P450 Engineering Database: a navigation and prediction tool for the cytochrome P450 protein family. *Bioinformatics* 2007, 23:2015-2017.
27. Sirim D, Wagner F, Lisitsa A, Pleiss J: The cytochrome P450 engineering database: Integration of biochemical properties. *BMC Biochem* 2009, 10:27.
28. Park J, Lee S, Choi J, Ahn K, Park B, Kang S, Lee YH: Fungal cytochrome P450 database. *BMC Genomics* 2008, 9:402.
29. Nelson DR: The cytochrome p450 homepage. *Hum Genomics* 2009, 4:59-65.

30. Poulos TL, Finzel BC, Gunsalus IC, Wagner GC, Kraut J: The 2.6-A crystal structure of *Pseudomonas putida* cytochrome P450. *J Biol Chem* 1985, 260:16122-16130.
31. Ravichandran KG, Boddupalli SS, Hasemann CA, Peterson JA, Deisenhofer J: Crystal structure of hemoprotein domain of P450$_{BM3}$, a prototype for microsomal P450s. *Science* 1993, 261:731-736.
32. Hasemann CA, Ravichandran KG, Peterson JA, Deisenhofer J: Crystal structure and refinement of cytochrome P450$_{terp}$ at 2.3 A resolution. *J Mol Biol* 1994, 236:1169-1185.
33. Cupp-Vickery JR, Poulos TL: Structure of cytochrome P450$_{eryF}$ involved in erythromycin biosynthesis. *Nat Struct Biol* 1995, 2:144-153.
34. Park SY, Shimizu H, Adachi S, Nakagawa A, Tanaka I, Nakahara K, Shoun H, Obayashi E, Nakamura H, Iizuka T, Shiro Y: Crystal structure of nitric oxide reductase from denitrifying fungus *Fusarium oxysporum*. *Nat Struct Biol* 1997, 4:827-832.
35. Meharenna YT, Li H, Hawkes DB, Pearson AG, De Voss J, Poulos TL: Crystal structure of P450$_{cin}$ in a complex with its substrate, 1,8-cineole, a close structural homologue to D-camphor, the substrate for P450$_{cam}$. *Biochemistry* 2004, 43:9487-9494.
36. Williams PA, Cosme J, Sridhar V, Johnson EF, McRee DE: Microsomal cytochrome P450 2C5: comparison to microbial P450s and unique features. *J Inorg Biochem* 2000, 81:183-190.
37. Williams PA, Cosme J, Ward A, Angove HC, Matak Vinkovic D, Jhoti H: Crystal structure of human cytochrome P450 2C9 with bound warfarin. *Nature* 2003, 424:464-468.
38. Wang JF, Zhang CC, Chou KC, Wei DQ: Structure of cytochrome P450s and personalized drug. *Curr Med Chem* 2009, 16:232-244.
39. Omura T, Sato R: The Carbon Monoxide-Binding Pigment of Liver Microsomes. I. Evidence for Its Hemoprotein Nature. *J Biol Chem* 1964, 239:2370-2378.
40. Bayer E, Hill HAO, Roder A, Williams RJ: Interaction between Haem-Iron and Thiols. *J Chem Soc Chem Commun* 1969:109.
41. Hill HAO, Roder A, Williams RJ: Haem-Protein Interaction in Cytochrome P450. *Biochem J* 1969, 115:59P-60P.
42. Poulos TL, Finzel BC, Howard AJ: Crystal structure of substrate-free *Pseudomonas putida* cytochrome P450. *Biochemistry* 1986, 25:5314-5322.
43. Poulos TL, Finzel BC, Howard AJ: High-resolution crystal structure of cytochrome P450$_{cam}$. *J Mol Biol* 1987, 195:687-700.
44. Graham SE, Peterson JA: How similar are P450s and what can their differences teach us? *Arch Biochem Biophys* 1999, 369:24-29.

45. Graham-Lorence S, Peterson JA: P450s: structural similarities and functional differences. *FASEB J* 1996, 10:206-214.
46. Li H, Poulos TL: Crystallization of cytochromes P450 and substrate-enzyme interactions. *Curr Top Med Chem* 2004, 4:1789-1802.
47. Isin EM, Guengerich FP: Substrate binding to cytochromes P450. *Anal Bioanal Chem* 2008, 392:1019-1030.
48. Schenkman JB, Sligar SG, Cinti DL: Substrate interaction with cytochrome P450. *Pharmacol Ther* 1981, 12:43-71.
49. Schenkman JB, Remmer H, Estabrook RW: Spectral studies of drug interaction with hepatic microsomal cytochrome. *Mol Pharmacol* 1967, 3:113-123.
50. Schenkman JB: Effects of Temperature and Substrates on Component Reactions of Hepatic Microsomal Mixed-Function Oxidase. *Molecular Pharmacology* 1972, 8:178-188.
51. Denisov IG, Makris TM, Sligar SG, Schlichting I: Structure and chemistry of cytochrome P450. *Chem Rev* 2005, 105:2253-2277.
52. Sligar SG: Coupling of spin, substrate, and redox equilibria in cytochrome P450. *Biochemistry* 1976, 15:5399-5406.
53. Filatov M, Harris N, Shaik S: On the "rebound" mechanism of alkane hydroxylation by cytochrome P450: Electronic structure of the intermediate and the electron transfer character in the rebound step. *Angew Chem Int Edit* 1999, 38:3510-3512.
54. Groves JT, Mcclusky GA: Aliphatic Hydroxylation Via Oxygen Rebound - Oxygen-Transfer Catalyzed by Iron. *J Am Chem Soc* 1976, 98:859-861.
55. Hamdane D, Zhang H, Hollenberg P: Oxygen activation by cytochrome P450 monooxygenase. *Photosynth Res* 2008, 98:657-666.
56. Davydov R, Makris TM, Kofman V, Werst DE, Sligar SG, Hoffman BM: Hydroxylation of camphor by reduced oxy-cytochrome $P450_{cam}$: mechanistic implications of EPR and ENDOR studies of catalytic intermediates in native and mutant enzymes. *J Am Chem Soc* 2001, 123:1403-1415.
57. Koppenol WH: Oxygen activation by cytochrome P450: a thermodynamic analysis. *J Am Chem Soc* 2007, 129:9686-9690.
58. Vaz AD, McGinnity DF, Coon MJ: Epoxidation of olefins by cytochrome P450: evidence from site-specific mutagenesis for hydroperoxo-iron as an electrophilic oxidant. *Proc Natl Acad Sci U S A* 1998, 95:3555-3560.
59. Jin S, Bryson TA, Dawson JH: Hydroperoxoferric heme intermediate as a second electrophilic oxidant in cytochrome P450-catalyzed reactions. *J Biol Inorg Chem* 2004, 9:644-653.

60. Sligar SG, Makris TM, Denisov IG: Thirty years of microbial P450 monooxygenase research: peroxo-heme intermediates--the central bus station in heme oxygenase catalysis. *Biochem Biophys Res Commun* 2005, 338:346-354.
61. Basch H, Mogi K, Musaev DG, Morokuma K: Mechanism of the methane -> methanol conversion reaction catalyzed by methane monooxygenase: A density functional study. *J Am Chem Soc* 1999, 121:7249-7256.
62. Jin Y, Lipscomb JD: Mechanistic insights into C-H activation from radical clock chemistry: oxidation of substituted methylcyclopropanes catalyzed by soluble methane monooxygenase from *Methylosinus trichosporium* OB3b. *Biochim Biophys Acta* 2000, 1543:47-59.
63. Kopp DA, Lippard SJ: Soluble methane monooxygenase: activation of dioxygen and methane. *Curr Opin Chem Biol* 2002, 6:568-576.
64. Itoh S: Mononuclear copper active-oxygen complexes. *Curr Opin Chem Biol* 2006, 10:115-122.
65. Cirino PC, Arnold FH: A Self-Sufficient Peroxide-Driven Hydroxylation Biocatalyst. *Angew Chem Int Ed Engl* 2003, 42:3299-3301.
66. Kotze AC: Peroxide-supported in-vitro cytochrome P450 activities in *Haemonchus contortus*. *Int J Parasitol* 1999, 29:389-396.
67. Rabe KS, Kiko K, Niemeyer CM: Characterization of the peroxidase activity of CYP119, a thermostable P450 from *Sulfolobus acidocaldarius*. *Chembiochem* 2008, 9:420-425.
68. Rabe KS, Spengler M, Erkelenz M, Muller J, Gandubert VJ, Hayen H, Niemeyer CM: Screening for cytochrome P450 reactivity by harnessing catalase as reporter enzyme. *Chembiochem* 2009, 10:751-757.
69. Matsunaga I, Shiro Y: Peroxide-utilizing biocatalysts: structural and functional diversity of heme-containing enzymes. *Curr Opin Chem Biol* 2004, 8:127-132.
70. Matsunaga I, Yokotani N, Gotoh O, Kusunose E, Yamada M, Ichihara K: Molecular cloning and expression of fatty acid alpha-hydroxylase from *Sphingomonas paucimobilis*. *J Biol Chem* 1997, 272:23592-23596.
71. Matsunaga I, Ueda A, Fujiwara N, Sumimoto T, Ichihara K: Characterization of the ybdT gene product of *Bacillus subtilis*: novel fatty acid beta-hydroxylating cytochrome P450. *Lipids* 1999, 34:841-846.
72. Sono M, Roach MP, Coulter ED, Dawson JH: Heme-containing oxygenases. *Chem Rev* 1996, 96:2841-2888.
73. Bernhardt R: Cytochromes P450 as versatile biocatalysts. *J Biotechnol* 2006, 124:128-145.

74. Cryle MJ, Stok JE, De Voss JJ: Reactions catalyzed by bacterial cytochromes P450. *Australian Journal of Chemistry* 2003, 56:749-762.
75. Guengerich FP: Uncommon P450-catalyzed reactions. *Curr Drug Metab* 2001, 2:93-115.
76. Guengerich FP: Mechanisms of cytochrome P450 substrate oxidation: MiniReview. *J Biochem Mol Toxicol* 2007, 21:163-168.
77. Urlacher VB, Eiben S: Cytochrome P450 monooxygenases: perspectives for synthetic application. *Trends Biotechnol* 2006, 24:324-330.
78. Hilker BL, Fukushige H, Hou C, Hildebrand D: Comparison of *Bacillus monooxygenase* genes for unique fatty acid production. *Prog Lipid Res* 2008, 47:1-14.
79. Miura Y, Fulco AJ: Omega-1, Omega-2 and Omega-3 hydroxylation of long-chain fatty acids, amides and alcohols by a soluble enzyme system from *Bacillus megaterium*. *Biochim Biophys Acta* 1975, 388:305-317.
80. Rheinwald JG, Chakrabarty AM, Gunsalus IC: A transmissible plasmid controlling camphor oxidation in *Pseudomonas putida*. *Proc Natl Acad Sci U S A* 1973, 70:885-889.
81. Tang L, Shah S, Chung L, Carney J, Katz L, Khosla C, Julien B: Cloning and heterologous expression of the epothilone gene cluster. *Science* 2000, 287:640-642.
82. Bruntner C, Lauer B, Schwarz W, Mohrle V, Bormann C: Molecular characterization of co-transcribed genes from *Streptomyces tendae* Tu901 involved in the biosynthesis of the peptidyl moiety of the peptidyl nucleoside antibiotic nikkomycin. *Mol Gen Genet* 1999, 262:102-114.
83. Kelly SL, Lamb DC, Kelly DE: Cytochrome P450 biodiversity and biotechnology. *Biochem Soc Trans* 2006, 34:1159-1160.
84. Lewis DF: *Cytochromes P450: Structure, Function and Mechanism, Vol 1*. London: Taylor & Francis; 1996.
85. Ortiz de Montellano PR (Ed.). Cytochrome P450: Structure, Mechanism, and Biochemistry 3rd edition. New York: Kluwer Academic/Plenum Press; 2005.
86. Torres Pazmino DE, Winkler M, Glieder A, Fraaije MW: Monooxygenases as biocatalysts: Classification, mechanistic aspects and biotechnological applications. *J Biotechnol* 2010, 146:9-24.
87. Urlacher VB: Catalysis with Cytochrome P450 Monooxygenases. In *Green Catalysis, Vol 3: Biocatalysis. Volume 3*. First edition. Edited by Crabtree RH. Weinheim: Wiley-VCH; 2009: 1-25 [Anastas PT (Series Editor): *Handbook of Green Chemistry*].
88. Werck-Reichhart D, Feyereisen R: Cytochromes P450: a success story. *Genome Biol* 2000, 1:REVIEWS3003.

89. Wong LL: Cytochrome P450 monooxygenases. *Curr Opin Chem Biol* 1998, 2:263-268.
90. van Beilen JB, Duetz WA, Schmid A, Witholt B: Practical issues in the application of oxygenases. *Trends Biotechnol* 2003, 21:170-177.
91. Petzoldt K, Annen K, Laurent H, Wiechert R: Process for the preparation of 11-β-hydroxy steroids. US: Schering Aktiengesellschaft (Berlin, Germany); 1982.
92. Hogg JA: Steroids, the steroid community, and Upjohn in perspective: a profile of innovation. *Steroids* 1992, 57:593-616.
93. Peterson DH, Murray HC, Eppstein SH, Reineke LM, Weintraub A, Meister PD, Leigh HM: Microbiological Transformations of Steroids. 1. Introduction of Oxygen at Carbon-11 of Progesterone. *J Am Chem Soc* 1952, 74:5933-5936.
94. Serizawa N: Development of two-step fermentation-based production of pravastatin, a HMG-CoA reductase. *Journal of Synthetic Organic Chemistry Japan* 1997, 55:334-338.
95. Serizawa N, Nakagawa K, Hamano K, Tsujita Y, Terahara A, Kuwano H: Microbial Hydroxylation of Ml-236b (Compactin) and Monacolin-K (Mb-530b). *J Antibiot (Tokyo)* 1983, 36:604-607.
96. Arsenault PR, Wobbe KK, Weathers PJ: Recent advances in artemisinin production through heterologous expression. *Curr Med Chem* 2008, 15:2886-2896.
97. Barnes HJ: Maximizing expression of eukaryotic cytochrome P450s in Escherichia coli. *Method Enzymol* 1996, 272:3-14.
98. Gonzalez FJ, Korzekwa KR: Cytochromes P450 expression systems. *Annu Rev Pharmacol Toxicol* 1995, 35:369-390.
99. Guengerich FP, Gillam EM, Shimada T: New applications of bacterial systems to problems in toxicology. *Crit Rev Toxicol* 1996, 26:551-583.
100. Park JB: Oxygenase-based whole-cell biocatalysis in organic synthesis. *J Microbiol Biotechnol* 2007, 17:379-392.
101. Yun CH, Yim SK, Kim DH, Ahn T: Functional expression of human cytochrome P450 enzymes in *Escherichia coli*. *Curr Drug Metab* 2006, 7:411-429.
102. Hannemann F, Bichet A, Ewen KM, Bernhardt R: Cytochrome P450 systems--biological variations of electron transport chains. *Biochim Biophys Acta* 2007, 1770:330-344.
103. Daiber A, Shoun H, Ullrich V: Nitric oxide reductase (P450$_{nor}$) from *Fusarium oxysporum*. *J Inorg Biochem* 2005, 99:185-193.
104. Wright RL, Harris K, Solow B, White RH, Kennelly PJ: Cloning of a potential cytochrome P450 from the archaeon *Sulfolobus solfataricus*. *FEBS Lett* 1996, 384:235-239.
105. Nishida CR, Ortiz de Montellano PR: Thermophilic cytochrome P450 enzymes. *Biochem Biophys Res Commun* 2005, 338:437-445.

106. Hawkes DB, Adams GW, Burlingame AL, Ortiz de Montellano PR, De Voss JJ: Cytochrome P450$_{cin}$ (CYP176A), isolation, expression, and characterization. *J Biol Chem* 2002, 277:27725-27732.
107. Stok JE, De Voss J: Expression, purification, and characterization of BioI: a carbon-carbon bond cleaving cytochrome P450 involved in biotin biosynthesis in *Bacillus subtilis*. *Arch Biochem Biophys* 2000, 384:351-360.
108. Lawson RJ, von Wachenfeldt C, Haq I, Perkins J, Munro AW: Expression and characterization of the two flavodoxin proteins of *Bacillus subtilis*, YkuN and YkuP: biophysical properties and interactions with cytochrome P450$_{BioI}$. *Biochemistry* 2004, 43:12390-12409.
109. Green AJ, Munro AW, Cheesman MR, Reid GA, von Wachenfeldt C, Chapman SK: Expression, purification and characterisation of a *Bacillus subtilis* ferredoxin: a potential electron transfer donor to cytochrome P450$_{BioI}$. *J Inorg Biochem* 2003, 93:92-99.
110. Porter TD, Kasper CB: NADPH-cytochrome P450 oxidoreductase: flavin mononucleotide and flavin adenine dinucleotide domains evolved from different flavoproteins. *Biochemistry* 1986, 25:1682-1687.
111. Sevrioukova I, Truan G, Peterson JA: The flavoprotein domain of P450$_{BM3}$: expression, purification, and properties of the flavin adenine dinucleotide- and flavin mononucleotide-binding subdomains. *Biochemistry* 1996, 35:7528-7535.
112. Oster T, Boddupalli SS, Peterson JA: Expression, purification, and properties of the flavoprotein domain of cytochrome P450$_{BM3}$. Evidence for the importance of the amino-terminal region for FMN binding. *J Biol Chem* 1991, 266:22718-22725.
113. Boddupalli SS, Oster T, Estabrook RW, Peterson JA: Reconstitution of the fatty acid hydroxylation function of cytochrome P450$_{BM3}$ utilizing its individual recombinant hemo- and flavoprotein domains. *J Biol Chem* 1992, 267:10375-10380.
114. Sevrioukova I, Truan G, Peterson JA: Reconstitution of the fatty acid hydroxylase activity of cytochrome P450$_{BM3}$ utilizing its functional domains. *Arch Biochem Biophys* 1997, 340:231-238.
115. Ingelman M, Bianchi V, Eklund H: The three-dimensional structure of flavodoxin reductase from *Escherichia coli* at 1.7 A resolution. *J Mol Biol* 1997, 268:147-157.
116. McIver L, Leadbeater C, Campopiano DJ, Baxter RL, Daff SN, Chapman SK, Munro AW: Characterisation of flavodoxin NADP$^+$ oxidoreductase and flavodoxin; key components of electron transfer in *Escherichia coli*. *Eur J Biochem* 1998, 257:577-585.
117. Sevrioukova IF, Li H, Poulos TL: Crystal structure of putidaredoxin reductase from *Pseudomonas putida*, the final structural component of the cytochrome P450$_{cam}$ monooxygenase. *J Mol Biol* 2004, 336:889-902.

118. Sevrioukova IF, Poulos TL, Churbanova IY: Crystal structure of the putidaredoxin reductase x putidaredoxin electron transfer complex. *J Biol Chem* 2010, 285:13616-13620.
119. Sevrioukova IF, Poulos TL: Arginines 65 and 310 in putidaredoxin reductase are critical for interaction with putidaredoxin. *Biochemistry* 2010, 49:5160-5166.
120. Bernhardt R: Cytochrome P450: structure, function, and generation of reactive oxygen species. *Rev Physiol Biochem Pharmacol* 1996, 127:137-221.
121. Lambeth JD, Seybert DW, Lancaster JR, Jr., Salerno JC, Kamin H: Steroidogenic electron transport in adrenal cortex mitochondria. *Mol Cell Biochem* 1982, 45:13-31.
122. Ziegler GA, Vonrhein C, Hanukoglu I, Schulz GE: The structure of adrenodoxin reductase of mitochondrial P450 systems: electron transfer for steroid biosynthesis. *J Mol Biol* 1999, 289:981-990.
123. Ziegler GA, Schulz GE: Crystal structures of adrenodoxin reductase in complex with $NADP^+$ and NADPH suggesting a mechanism for the electron transfer of an enzyme family. *Biochemistry* 2000, 39:10986-10995.
124. Massey V: Introduction: flavoprotein structure and mechanism. *FASEB J* 1995, 9:473-475.
125. Bruschi M, Guerlesquin F: Structure, function and evolution of bacterial ferredoxins. *FEMS Microbiol Rev* 1988, 4:155-175.
126. Ewen KM, Kleser M, Bernhardt R: Adrenodoxin: The archetype of vertebrate-type [2Fe-2S] cluster ferredoxins. *Biochim Biophys Acta* 2010.
127. Degtyarenko KN, Kulikova TA: Evolution of bioinorganic motifs in P450-containing systems. *Biochem Soc Trans* 2001, 29:139-147.
128. Purdy MM, Koo LS, Ortiz de Montellano PR, Klinman JP: Steady-state kinetic investigation of cytochrome $P450_{cam}$: interaction with redox partners and reaction with molecular oxygen. *Biochemistry* 2004, 43:271-281.
129. Sibbesen O, De Voss JJ, Montellano PR: Putidaredoxin reductase-putidaredoxin-cytochrome $P450_{cam}$ triple fusion protein. Construction of a self-sufficient *Escherichia coli* catalytic system. *J Biol Chem* 1996, 271:22462-22469.
130. Kim D, Ortiz de Montellano PR: Tricistronic overexpression of cytochrome $P450_{cam}$, putidaredoxin, and putidaredoxin reductase provides a useful cell-based catalytic system. *Biotechnol Lett* 2009.
131. Pikuleva IA, Tesh K, Waterman MR, Kim Y: The tertiary structure of full-length bovine adrenodoxin suggests functional dimers. *Arch Biochem Biophys* 2000, 373:44-55.
132. Smillie RM: Isolation of Phytoflavin, A Flavoprotein with Chloroplast Ferredoxin Activity. *Plant Physiol* 1965, 40:1124-1128.

133. Smillie RM: Isolation of two proteins with chloroplast ferredoxin activity from a blue-green alga. *Biochem Biophys Res Commun* 1965, 20:621-629.
134. Knight E, Jr., D'Eustachio AJ, Hardy RW: Flavodoxin: a flavoprotein with ferredoxin activity from *Clostrium pasteurianum*. *Biochim Biophys Acta* 1966, 113:626-628.
135. Sancho J: Flavodoxins: sequence, folding, binding, function and beyond. *Cell Mol Life Sci* 2006, 63:855-864.
136. Hall DA, Vander Kooi CW, Stasik CN, Stevens SY, Zuiderweg ER, Matthews RG: Mapping the interactions between flavodoxin and its physiological partners flavodoxin reductase and cobalamin-dependent methionine synthase. *Proc Natl Acad Sci U S A* 2001, 98:9521-9526.
137. Hawkes DB, Slessor KE, Bernhardt PV, De Voss JJ: Cloning, expression and purification of cindoxin, an unusual Fmn-containing cytochrome P450 redox partner. *Chembiochem* 2010, 11:1107-1114.
138. Momoi K, Hofmann U, Schmid RD, Urlacher VB: Reconstitution of β-carotene hydroxylase activity of thermostable CYP175A1 monooxygenase. *Biochem Biophys Res Commun* 2006, 339:331-336.
139. Jenkins CM, Waterman MR: Flavodoxin and NADPH-flavodoxin reductase from *Escherichia coli* support bovine cytochrome P450c17 hydroxylase activities. *J Biol Chem* 1994, 269:27401-27408.
140. Girhard M, Machida K, Itoh M, Schmid RD, Arisawa A, Urlacher VB: Regioselective biooxidation of (+)-valencene by recombinant *E. coli* expressing CYP109B1 from *Bacillus subtilis* in a two-liquid-phase system. *Microb Cell Fact* 2009, 8:36.
141. Arisawa A, Agematu H: A Modular Approach to Biotransformation Using Microbial Cytochrome P450 Monooxygenases. In *Modern Biooxidation*. First edition. Edited by Schmid RD, Urlacher VB. Weinheim: Wiley-VCH; 2007: 177-192.
142. Agematu H, Matsumoto N, Fujii Y, Kabumoto H, Doi S, Machida K, Ishikawa J, Arisawa A: Hydroxylation of testosterone by bacterial cytochromes P450 using the *Escherichia coli* expression system. *Biosci Biotechnol Biochem* 2006, 70:307-311.
143. Yasutake Y, Imoto N, Fujii Y, Fujii T, Arisawa A, Tamura T: Crystal structure of cytochrome P450 MoxA from *Nonomuraea recticatena* (CYP105). *Biochem Biophys Res Commun* 2007, 361:876-882.
144. Girhard M, Schuster S, Dietrich M, Durre P, Urlacher VB: Cytochrome P450 monooxygenase from *Clostridium acetobutylicum*: A new α-fatty acid hydroxylase. *Biochem Biophys Res Commun* 2007, 362:114-119.
145. Nölling J, Breton G, Omelchenko MV, Makarova KS, Zeng Q, Gibson R, Lee HM, Dubois J, Qiu D, Hitti J, et al: Genome sequence and comparative analysis of the

solvent-producing bacterium *Clostridium acetobutylicum*. *J Bacteriol* 2001, 183:4823-4838.

146. Shoji O, Fujishiro T, Nagano S, Tanaka S, Hirose T, Shiro Y, Watanabe Y: Understanding substrate misrecognition of hydrogen peroxide dependent cytochrome P450 from *Bacillus subtilis*. *J Biol Inorg Chem* 2010.

147. Lee DS, Yamada A, Sugimoto H, Matsunaga I, Ogura H, Ichihara K, Adachi S, Park SY, Shiro Y: Substrate recognition and molecular mechanism of fatty acid hydroxylation by cytochrome P450 from *Bacillus subtilis*. Crystallographic, spectroscopic, and mutational studies. *J Biol Chem* 2003, 278:9761-9767.

148. Girhard M, Klaus T, Khatri Y, Bernhardt R, Urlacher VB: Characterization of the versatile monooxygenase CYP109B1 from *Bacillus subtilis*. *Appl Microbiol Biotechnol* 2010, 87:595-607.

149. Miura Y, Fulco AJ: w-1, w-2 and w-3 hydroxylation of long-chain fatty acids, amides and alcohols by a soluble enzyme system from *Bacillus megaterium*. *Biochimica et Biophysica Acta* 1975, 388:305-317.

150. Brenna E, Fuganti C, Serra S, Kraft P: Optically active ionones and derivatives: Preparation and olfactory properties. *European Journal of Organic Chemistry* 2002:967-978.

151. Eschenmoser W, Uebelhart P, Eugster CH: Synthesis and Chirality of the Enantiomeric 6-Hydroxy-α-Ionones and of *Cis-* and *Trans*-5,6-Dihydroxy-5,6-Dihydro-β-Ionones. *Helvetica Chimica Acta* 1981, 64:2681-2690.

152. Larroche C, Creuly C, Gros JB: Fed-Batch Biotransformation of β-Ionone by *Aspergillus niger*. *Applied Microbiology and Biotechnology* 1995, 43:222-227.

153. Khatri Y, Girhard M, Romankiewicz A, Ringle M, Hannemann F, Urlacher VB, Hutter MC, Bernhardt R: Regioselective hydroxylation of norisoprenoids by CYP109D1 from *Sorangium cellulosum* So ce56. *Appl Microbiol Biotechnol* 2010, 88:485-495.

154. Altschul SF, Madden TL, Schaffer AA, Zhang J, Zhang Z, Miller W, Lipman DJ: Gapped BLAST and PSI-BLAST: a new generation of protein database search programs. *Nucleic Acids Res* 1997, 25:3389-3402.

155. Urlacher VB, Makhsumkhanov A, Schmid RD: Biotransformation of β-ionone by engineered cytochrome P450$_{BM3}$. *Appl Microbiol Biotechnol* 2006, 70:53-59.

156. Celik A, Flitsch SL, Turner NJ: Efficient terpene hydroxylation catalysts based upon P450 enzymes derived from actinomycetes. *Org Biomol Chem* 2005, 3:2930-2934.

157. Kaya K, Ramesha CS, Thompson GA, Jr.: On the formation of α-hydroxy fatty acids. Evidence for a direct hydroxylation of nonhydroxy fatty acid-containing sphingolipids. *J Biol Chem* 1984, 259:3548-3553.

158. Koch AK, Kappeli O, Fiechter A, Reiser J: Hydrocarbon assimilation and biosurfactant production in *Pseudomonas aeruginosa* mutants. *J Bacteriol* 1991, 173:4212-4219.
159. Shoji O, Fujishiro T, Nakajima H, Kim M, Nagano S, Shiro Y, Watanabe Y: Hydrogen peroxide dependent monooxygenations by tricking the substrate recognition of cytochrome P450$_{Bs\beta}$. *Angew Chem Int Ed Engl* 2007, 46:3656-3659.
160. Chefson A, Auclair K: Progress towards the easier use of P450 enzymes. *Mol Biosyst* 2006, 2:462-469.
161. Kellner DG, Maves SA, Sligar SG: Engineering cytochrome P450s for bioremediation. *Curr Opin Biotechnol* 1997, 8:274-278.
162. Rosic NN, Huang W, Johnston WA, DeVoss JJ, Gillam EM: Extending the diversity of cytochrome P450 enzymes by DNA family shuffling. *Gene* 2007, 395:40-48.
163. Whitehouse CJ, Bell SG, Tufton HG, Kenny RJ, Ogilvie LC, Wong LL: Evolved CYP102A1 (P450$_{BM3}$) variants oxidise a range of non-natural substrates and offer new selectivity options. *Chem Commun (Camb)* 2008:966-968.
164. Wong TS, Tee KL, Hauer B, Schwaneberg U: Sequence saturation mutagenesis (SeSaM): a novel method for directed evolution. *Nucleic Acids Res* 2004, 32:e26.
165. Bell SG, Dale A, Rees NH, Wong LL: A cytochrome P450 class I electron transfer system from *Novosphingobium aromaticivorans*. *Appl Microbiol Biotechnol* 2010, 86:163-175.
166. Bell SG, Wong LL: P450 enzymes from the bacterium *Novosphingobium aromaticivorans*. *Biochem Biophys Res Commun* 2007, 360:666-672.
167. Honda Malca S, Girhard M, Schuster S, Dürre P, Urlacher VB: Expression, purification and characterization of two *Clostridium acetobutylicum* flavodoxins: Potential electron transfer partners for CYP152A2. *Biochim Biophys Acta* 2010:in press.
168. Kunst F, Ogasawara N, Moszer I, Albertini AM, Alloni G, Azevedo V, Bertero MG, Bessieres P, Bolotin A, Borchert S, et al: The complete genome sequence of the gram-positive bacterium *Bacillus subtilis*. *Nature* 1997, 390:249-256.
169. Hannemann F, Virus C, Bernhardt R: Design of an Escherichia coli system for whole cell mediated steroid synthesis and molecular evolution of steroid hydroxylases. *J Biotechnol* 2006, 124:172-181.
170. Zehentgruber D, Hannemann F, Bleif S, Bernhardt R, Lutz S: Towards preparative scale steroid hydroxylation with cytochrome P450 monooxygenase CYP106A2. *Chembiochem* 2010, 11:713-721.
171. Ullah AJ, Murray RI, Bhattacharyya PK, Wagner GC, Gunsalus IC: Protein components of a cytochrome P450 linalool 8-methyl hydroxylase. *J Biol Chem* 1990, 265:1345-1351.

172. Mandai T, Fujiwara S, Imaoka S: A novel electron transport system for thermostable CYP175A1 from *Thermus thermophilus* HB27. *Febs J* 2009, 276:2416-2429.
173. Furusawa M, Hashimoto T, Noma Y, Asakawa Y: Highly efficient production of nootkatone, the grapefruit aroma from valencene, by biotransformation. *Chem Pharm Bull (Tokyo)* 2005, 53:1513-1514.
174. Sakamaki H, Itoh K, Taniai T, Kitanaka S, Takagi Y, Chai W, Horiuchi CA: Biotransformation of valencene by cultured cells of *Gynostemma pentaphyllum*. *Journal of Molecular Catalysis B-Enzymatic* 2005, 32:103-106.
175. Okuda M, Sonohara H, Takigawa H, Tajima K, Ito S: Nootkatone manufacture with *Rhodococcus* from valencene. Japan, 1994.
176. Fraatz MA, Riemer SJL, Stöber R, R. K, Nimtz M, Berger RG, Zorn H: A novel oxygenase from *Pleurotus sapidus* transforms valencene to nootkatone. *J Mol Catal B Enzym* 2009.

5. Appendix

Appendix 1: List of abbreviations

°C	degree Celsius
a	year (anno)
Å	ångström
Abs	absorption
AdR	bovine adrenodoxin reductase
Adx	bovine adrenodoxin
Amp	ampicillin
approx.	approximately
APS	ammonium persulfate
bp	base pair(s)
BMR	CYP102A1-diflavin reductase domain
B. subtilis	Bacillus subtilis
B. megaterium	Bacillus megaterium
BMP	hemoproteins domain of $P450_{BM3}$
BMR	reductase domain of $P450_{BM3}$
C. acetobutylicum	Clostridium acetobutylicum
C. braakii	Citrobacter braakii
C-terminal	carboxy-terminal
CYP	cytochrome P450 monooxygenase
d	day
Da	dalton
DMAPP	dimethylallyl diphosphate
DNA	desoxyribonucleic acid
E.C.	enzyme class
E. coli	Escherichia coli
e.g.	for example
FAD	flavin adenine dinucleotide
FdR	E. coli flavodoxin reductase
Fdx	E. coli flavodoxin
FMN	flavin mononucleotide
g	gram
h	hour

HCl	hydrochloric acid
IPP	isopentenyl diphosphate
IPTG	isopropylthio-β-D-galactoside
Kan	kanamycin
kDa	kilodalton
M	molar
m	milli
mg	milligram
min	minute
ml	millilitre
mM	millimolar
mV	millivolt
µ	micro
n	nano
NAD(P)H	β-nicotinamide adenine dinucleotide (phosphate)
nm	nanometre
N-terminal	amino-terminal
OD_{600}	optical density at 600 nm
P. putida	*Pseudomonas putida*
P450	cytochrome P450 monooxygenase
PCR	Polymerase chain reaction
PdR	putidaredoxin reductase
Pdx	putidaredoxin
SDS	sodium dodecyl sulfate
SDS-PAGE	SDS-Polyacrylamid gelelectrophoresis
sec	second
SRS	substrate recognition site
T	temperature
t	time
Tris	Tris (hydroxymethyl) aminomethane
U	unit
YkuN	*B. subtilis* flavodoxin N
YkuP	*B. subtilis* flavodoxin P

Appendix 2: Table of amino acid abbreviations

Table 5-1: Amino acid abbreviations

Amino acid	3 Letter	1 Letter	Amino acid	3 Letter	1 Letter
Alanine	Ala	A	Leucine	Leu	L
Arginine	Arg	R	Lysine	Lys	K
Asparagine	Asn	N	Methionine	Met	M
Aspartic acid	Asp	D	Phenylalanine	Phe	F
Cysteine	Cys	C	Proline	Pro	P
Glutamic acid	Glu	E	Serine	Ser	S
Glutamine	Gln	Q	Threonine	Thr	T
Glycine	Gly	G	Tryptophane	Trp	W
Histidine	His	H	Tyrosine	Tyr	Y
Isoleucine	Ile	I	Valine	Val	V

Die VDM Verlagsservicegesellschaft sucht für wissenschaftliche Verlage abgeschlossene und herausragende

Dissertationen, Habilitationen, Diplomarbeiten, Master Theses, Magisterarbeiten usw.

für die kostenlose Publikation als Fachbuch.

Sie verfügen über eine Arbeit, die hohen inhaltlichen und formalen Ansprüchen genügt, und haben Interesse an einer honorarvergüteten Publikation?

Dann senden Sie bitte erste Informationen über sich und Ihre Arbeit per Email an *info@vdm-vsg.de*.

Sie erhalten kurzfristig unser Feedback!

VDM Verlagsservicegesellschaft mbH
Dudweiler Landstr. 99   Telefon +49 681 3720 174
D - 66123 Saarbrücken   Fax   +49 681 3720 1749
www.vdm-vsg.de

Die VDM Verlagsservicegesellschaft mbH vertritt

Printed by Books on Demand GmbH, Norderstedt / Germany